T0296152

MICROSCOPIC ANALYSIS OF CATTLE-FOODS

MICROSCOPIC ANALYSIS OF CATTLE-FOODS

BY

T. N. MORRIS, B.A.

ST JOHN'S COLLEGE, CAMBRIDGE

Cambridge:

at the University Press

1917

CAMBRIDGE
UNIVERSITY PRESS

University Printing House, Cambridge CB2 8BS, United Kingdom

Cambridge University Press is part of the University of Cambridge.

It furthers the University's mission by disseminating knowledge in the pursuit of education, learning and research at the highest international levels of excellence.

www.cambridge.org
Information on this title: www.cambridge.org/9781107560048

First published 1917
First paperback edition 2015

A catalogue record for this publication is available from the British Library

ISBN 978-1-107-56004-8 Paperback

PREFACE

THIS book is intended as a brief guide in the recognition of the common legitimate constituents of cattle-foods, there being nothing in English dealing solely with this subject.

My thanks are due to Prof. Wood, Prof. Biffen, and Mr L. F. Newman of the School of Agriculture, Cambridge, for assistance and criticism.

T. N. M.

April, 1917

CONTENTS

PART I

PART II. HISTOLOGY

PART I

CHAPTER I

COMMON INGREDIENTS AND METHODS OF EXAMINATION OF CATTLE-FOODS

Concentrated cattle-foods are on the market largely in the form of cattle cakes which may contain from one to as many as nine or ten different vegetable constituents.

Cakes containing one constituent only, may be made from any one of the following seeds or fruits: Linseed, Cotton-seed, Rape, Soya beans, Palm nuts, Coco nuts, Ground nuts (also called Earth nuts, Pea nuts or Monkey nuts) and more rarely from Niger, Sunflower, Hemp, and other oil seeds.

Cakes such as 'Soycot' contain as the name indicates soya and cotton only. One may however find such mixtures as the following:

1	2	3
Cotton	Cotton	Cotton
Rape	Linseed	Soya
Rice meal	Carob bean	Palm nut
Soya beans	Rice meal	Locust (Carob) bean
Palm nut	Maize meal	Fenugreek
Pea nut	Pea nut	
Carob bean		
Fenugreek		

4

Cotton
Rape and other crucifers
Rice meal
Ground nut
Palm nut
Carob bean

5

Cotton
Rape
Rice meal
Palm nut
Linseed (trace)
Fenugreek

6

Cotton
Linseed
Carob bean
Wheat bran
Rice meal
Ground nut
Malt culms
Fenugreek

7

Cotton
Rice meal
Ground nut
Rape and other crucifers
Niger seed (trace)
Soya
Wheat bran

8

Cotton
Ground nut
Rice meal
Wheat bran

9

Cotton
Rape
Carob bean
Pea

10

Cotton
Rape
Soya
Rice

The following materials, not necessarily all present together, have been found in pig-meals:

Palm nut, Hemp seed, Soya bean, Wheat bran, Rice meal, Rice middlings, Treacle, and various vegetable pulps to absorb the treacle.

The chief ingredient of 'Cream Equivalents' and Calf meals is usually Linseed, but Maize flour, Maize germ meal, Palm nut meal, Soya bean meal, Rice meal, Rape, and Carob beans have also been found in them, with Fenugreek as a flavouring material.

The feeding cakes examined have on the whole been free from weed seeds. Black bindweed was recognised in one instance in very small quantity, but the very characteristic seed-coats of corn-cockle were not found at all.

A certain amount of sand may be present owing to imperfect cleaning of the ingredients—especially ground nut.

Identification of materials.

The various constituents of food-stuffs are identified chiefly by means of hard and resistant parts such as seed-coats, fruit walls, the chaff or paleae (in the case of cereals), and by cell-constituents such as starch grains, aleurone grains, crystals, etc.

It is necessary then, before examining the food-stuffs directly, to make a thorough study of raw materials in the shape of seeds, fruits, or other parts of plants, known to be used in their manufacture, or likely to be used owing to their cheapness or feeding value.

Methods of examining raw materials.

Seed-coats, paleae, etc., should be examined in cross-section and in surface view.

Sectioning. If the seeds are not too hard, the best cross-sections are obtained without soaking them in water. Cruciferous and other seeds containing mucilaginous material, should be cut dry or with the razor wetted with alcohol, and mounted in alcohol or strong glycerine. If they are mounted in alcohol, the effect of running water under the cover-glass can be studied and is often instructive.

If the seeds are soaked in water before being sectioned they should not be allowed to get too soft, as it is difficult to cut thin sections of very soft tissues.

Many of the seeds used in cakes are large enough to be held in the fingers when being sectioned. Very small seeds can be held between pieces of elder-pith, or if necessary they may be embedded in solid paraffin and so held. Even materials softened in water can be embedded in solid paraffin if the surface is carefully dried. The object to be cut is introduced into a cavity in a stick of paraffin and the paraffin is melted round it with a hot wire.

Two razors should be kept, one hollow ground for fine work, and one with a bevelled edge for hard tissues.

As a rule it is not necessary to go through the long process of impregnating with paraffin and embedding for microtome work unless it is desired to settle very minute points of structure. For all practical purposes hand sections only are required.

Surface preparations. These are of the greatest importance from the point of view of diagnosis, since surface preparations, more or less torn and ragged, are what the food microscopist has to recognise.

For making surface preparations the seeds must be soaked in water until soft. The various layers of the coats can then be separated from one another by scraping with a scalpel or with a fine needle, or thin surface sections can be cut by means of a razor. The appearance of the seed-coat in its entire thickness should also be carefully studied.

It is a good plan, in mounting a surface preparation to tear it in halves and mount one piece with the outer and one with the inner surface uppermost, noting which is which.

The position and thickness of the various layers may be arrived at by focussing with a graduated fine adjustment, or better, by measuring the layers in section with the help of an eye-piece-micrometer scale.

Clearing. As a rule it is found that preparations are rendered too transparent and lose too much detail if they are cleared with Clove oil and mounted in Canada balsam.

The most useful clearing reagent is dilute potash (1·25 per cent.). Preparations are boiled in this for two or three minutes, and then in water to remove the potash; or if too delicate to stand boiling, they may be soaked in cold potash.

If mucilage or starch is present in large quantity the preparations may be boiled in very dilute hydrochloric or sulphuric acids before boiling in potash.

Maceration. For softening very hard tissues such as nut shells, etc., the preparations can be soaked for a time in Schulze's macerating fluid consisting of concentrated nitric acid with a few crystals of potassium chlorate added.

Mountants. The chief mountant for mere purposes of diagnosis is water. For permanent preparations glycerine or glycerine jelly should be used. Glycerine jelly tinted with gentian violet is also good and acts as a stain and mountant at the same time.

Glycerine jelly (Kaiser's) is made up as follows: soak 1 part of finest French gelatine 2 hours in 6 parts of distilled water. Add 7 parts of glycerine, and to each 100 grams of the mixture, 1 gram of strongest carbolic acid. Warm for 10–15 minutes with constant stirring until the flakes from the carbolic acid disappear. Filter through previously moistened glass wool. Warm as needed and remove with a glass rod[1].

Glycerine preparations may be ringed with Brunswick black. In doing this care must be taken to use exactly the right sized drop of glycerine as there must be none exuding from beneath the coverslip.

Preparations appear more transparent in glycerine than they do in water, and it is necessary that one should be able to recognise them in both mountants.

Staining. All diagnostic work can be carried on by means of unstained preparations; staining is only useful in bringing out more clearly fine points of structure.

The chief stains in use are Safranin (dilute watery solution), Gentian violet (dilute watery solution), or haematoxylin in watery solution. The sections can be soaked in these until coloured sufficiently; excess of stain can be removed by means of alcohol.

Chlor-zinc-iodine solution colours cellulose blue and lignified, suberized, and cuticularised tissues yellow; treatment with iodine and then with strong sulphuric acid has much the same effect.

Chlor-zinc-iodine is made up as follows: treat an excess of zinc with hydrochloric acid, evaporate to a specific gravity of 1·8 and filter through asbestos. As needed, saturate a small quantity of the sirupy liquid first with potassium iodide and finally with iodine.

The solution may also be prepared by dissolving 30 grams of zinc chloride, 5 grams of potassium iodide and 0·89 gram of iodine in 14 c.c. of water. The

[1] Winton, *Microscopy of Vegetable Foods*, p. 9.

solution should be freshly prepared and kept in a dark place[1].

Starch and Aleurone grains. A solution of iodine in potassium iodide, as is well-known, colours starch grains blue and aleurone grains, yellow. Starch grains should be mounted and examined first in water and then in iodine solution.

EXAMINATION OF PREPARED FOOD-STUFFS.

A. Preliminary. A preliminary examination of the cake or other food-stuff should be made with the naked eye or with a hand lens, as it is possible that some of the larger fragments such as pieces of carob bean pods, split carob beans, cotton-seed husks, broken rice grains, and perhaps rice paleae can be identified in this way.

Flavouring matters such as *fenugreek*, *turmeric*, *salt*, and *treacle*, can be recognised in the original material by their smell or taste.

B. Microscopic examination. The material is crushed in a mortar and portions of the powder are examined as follows:

(*a*) Mount some in water on a slide, and cover with a slip which should be pressed down and rubbed against the slide slightly. Examine under $\frac{2}{3}''$ and $\frac{1}{6}''$ objectives. Look for starch grains, brown wrinkled bodies of carob bean pods[2], dagger cells of rice paleae[3]. If brown bodies are present resembling those of the carob bean, run potash under the cover-glass, or mount some of the powder in potash and examine first cold and then

[1] Winton, *Microscopy of Vegetable Foods*, p. 8.

[2] See page 42 [3] See page 23.

after warming the slide over a flame. Dilute potash colours the brown bodies violet when warmed; strong potash colours them deep blue.

(*b*) Mount some more of the powder in a dilute solution of iodine in potassium iodide and examine again for starch and aleurone grains. If starch grains are present note their size and shape.

(*c*) Set aside a little of the powder to soften in water, so that softened but comparatively uninjured fragments may be at hand if required.

(*d*) Boil a few grammes of the powder in water; this swells the starch grains and dissolves much of the mucilage if any is present. If much starch or mucilage is present, the material should be boiled with very dilute hydrochloric or sulphuric acid. After the first boiling allow the fragments to settle, pour off as much of the liquid as possible and boil for two or three minutes with caustic potash or caustic soda (about 1·25 per cent.) to clear the tissues[1].

Again allow the fragments to settle, pour off as before and then boil with water to clear away the potash. A certain amount of separation of the fragments may be carried on at this stage by washing the material several times with water and pouring off the water each time before the lighter fragments have settled. The larger ones are thus obtained in a cleaner state, being less covered up with muddy-looking sediment. The liquid containing the lighter fragments should of course be kept for examination.

(*e*) The fragments, together with enough liquid to float them are then poured into a flat white dish or

[1] The alkali causes the cell-walls to swell somewhat and dissolves some of the proteins, colouring matters and other cell contents.

into a clock-glass which can be held over a white surface. Many of them can be recognised at this stage with some certainty by the naked eye. Samples should however be mounted on slides and examined under the $\frac{2}{3}''$ objective with No. 2 eye piece. Most of the fragments will be recognised by this power of the microscope, but the $\frac{1}{6}''$ objective should be turned on to decide doubtful points. A mechanical stage will be found to be very useful in working through slides.

PART II

HISTOLOGY

Introductory. The following account deals with the seeds and other parts of plants commonest in cattle-foods at the present time. For a more detailed account, which includes many more types than are dealt with here, the reader is referred to Winton's *Microscopy of Vegetable Foods*.

It is assumed that the reader is acquainted with the general morphology of fruits and seeds and with the essentials of plant histology. If he is not, he should refer to text-books on Botany.

It may be remarked however that the somewhat unusual term 'Spermoderm' is employed here as in Winton's work referred to above, to denote the true seed-coat derived from the two integuments of the ovule.

Terms such as pericarp, epicarp, mesocarp, endocarp relate to tissues formed from the ovary wall, and structures which possess these layers must be regarded as fruits and not seeds.

Thus the so-called seed or 'berry' of a cereal must be regarded as a fruit containing a single seed. It possesses typically an outer, often thick-walled epicarp, a mesocarp of somewhat similar cells with thinner walls, and an endocarp represented by 'cross cells' and 'tube cells.' These layers taken together make up the pericarp. The spermoderm is here represented merely by two very delicate layers of cells, although in some fruits and seeds it may consist of several layers and be of considerable thickness.

CHAPTER II

CEREALS

Wheat. (Triticum sativum.)

Wheat is often found in cattle-foods and cakes either crushed, or in the form of bran. No chaff is present as the grain threshes out clean, hence we need only consider here the structure of the bran coats and the starch grains. The bran coats consist of the following layers.

i. *Epicarp.* This forms a layer of colourless cells longitudinally elongated, except at the ends of the grain, and with fairly thick, beaded walls. Unicellular, thick walled, pointed hairs up to 1 mm. long are borne upon the epicarp at the apex of the grain. [See figs. 1, 2, and 3.]

ii. A *Mesocarp* of two or three layers of cells with beaded walls, similar to those of the epicarp. [See fig. 3.]

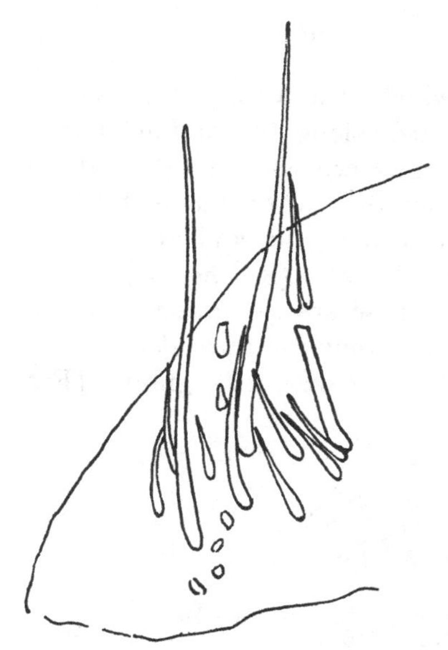

Fig. 1. Hairs on epicarp of wheat. ×66.

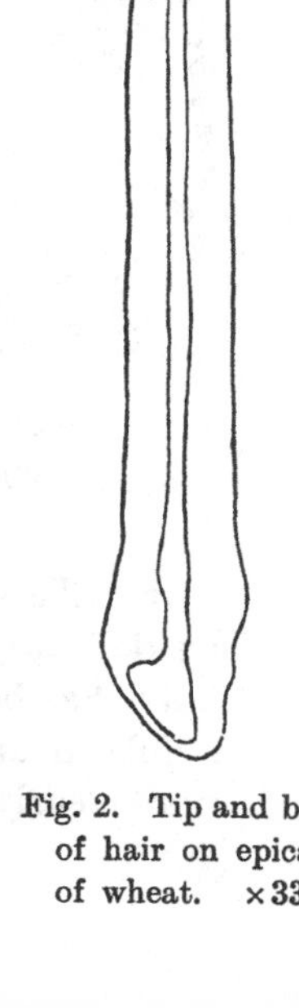

Fig. 2. Tip and base of hair on epicarp of wheat. ×333

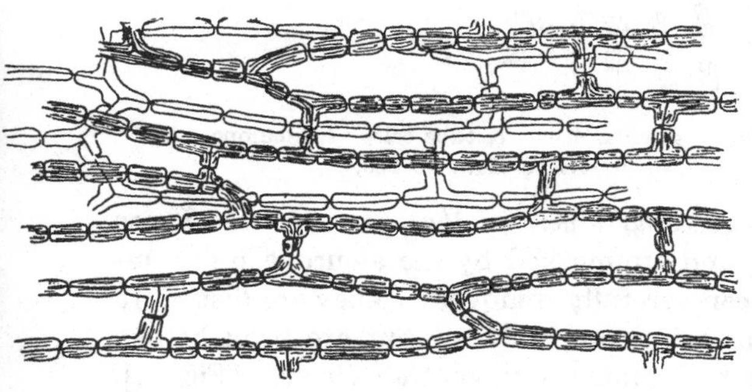

Fig. 3. *Wheat.* Cells of epicarp (shaded). One layer of mesocarp cells (unshaded) is also shown. Surface view. ×250.

iii. *Cross cells* which form a single layer of characteristic cells, transversely elongated, and arranged more or less regularly in rows according to the part of the grain under examination. They have fairly thick walls, beaded all the way round, and are up to 100 μ in length by 14–23 μ broad. [See fig. 4.]

iv. *Tube cells.* These are elongated, sinuous cells running longitudinally; they are present in great numbers, especially in certain parts of the grain. [Fig. 4.]

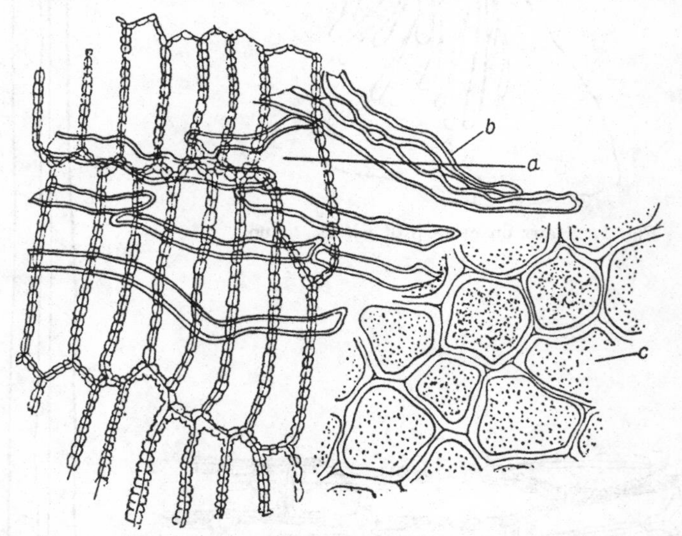

Fig. 4. *Wheat.* *a,* cross cells. *b,* tube cells. *c,* aleurone cells. All in surface view. ×250.

v. *Two crossing layers of the spermoderm.* These are delicate, and are masked by the aleurone layer unless it has been carefully removed. They are distinctly visible in unstained preparations, but are brought out more clearly by staining with gentian violet. [Fig. 5.]

vi. *Perisperm.* A layer of polygonal, flattened cells, not seen in unstained preparations. [Fig. 5.]

vii. *Aleurone layer of Endosperm.* A single layer of thick walled cells, polygonal in surface view and rectangular in section, with darkly granular contents which becomes considerably cleared in glycerine. This layer is the most prominent of all, and is apt to mask the other layers. In fig. 4 the cells are somewhat swollen with potash.

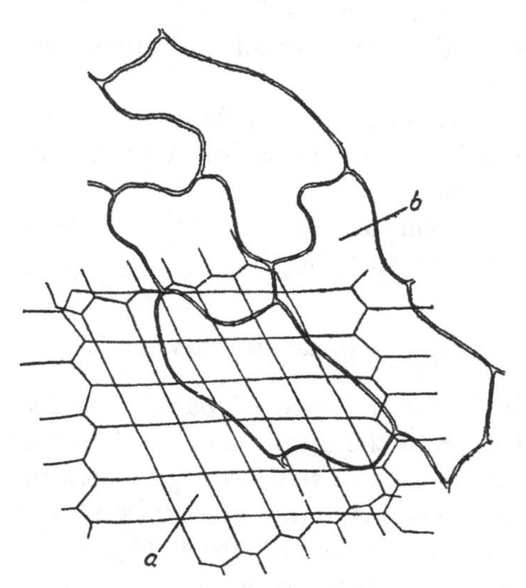

Fig. 5. *Wheat.* *a*, surface view of two crossing layers of spermoderm. *b*, perisperm. ×250. The cell-walls of the perisperm swell up in potash; here they are shown in an unswollen condition.

The starch grains are spherical and in two sizes, one 5, 6, to 10 μ, the other 20 to nearly 50 μ.

The large grains are smaller than those of rye and larger than those of barley. [See fig. 54.]

The distinguishing features in the case of wheat are as follows:

1. Fairly thick and *beaded* walls both in epicarp and mesocarp.

2. Hairs confined to apex of grain unlike those of oats. They are somewhat stouter than those of rye and have their ends more pointed.

3. Cross cells in a single layer [cf. barley which has a double layer]. Walls beaded all the way round unlike those of rye (compare figures).

4. Tube cells more distinct and numerous than in the other cereals.

5. The layers of the spermoderm cross one another, thus differing from those of barley in which they run in the same direction.

6. Single layer of aleurone cells (compare barley which has two or three layers).

Barley. (Hordeum sativum L.)

i. *Paleae*. In the case of barley the paleae are closely adherent to the grain, hence they must be considered in an account of the histological features of the grain. They have an outer epidermis of wavy cells from 50 μ to 120 μ long and from 10 μ to 16 μ wide (lumens 1 μ to 3 μ), together with round cells and twin cells. [See fig. 6.]

Fibres are present beneath the epidermis; next come several layers of a characteristic spongy parenchyma (see fig. 7) very distinct from the corresponding stellate spongy parenchyma found in oats. The inner epidermis is of delicate elongated cells provided with stomata and with short delicate hairs most numerous

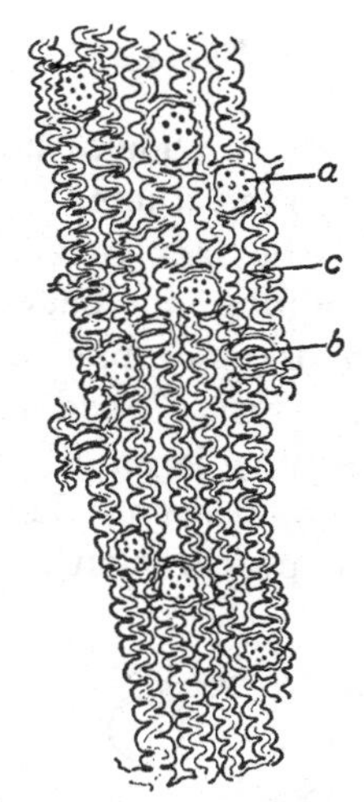

Fig. 6. *Barley.* Outer epidermis of palea. *a*, round cells. *b*, twin cells. *c*, wavy cells. ×250.

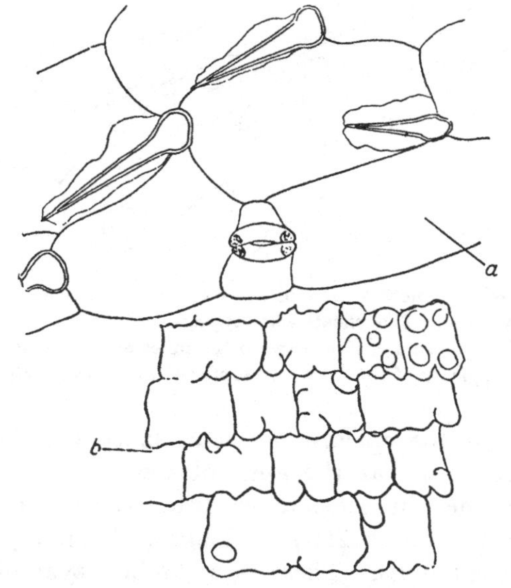

Fig. 7. *Barley.* *a*, inner epidermis of palea with hairs. *b*, spongy parenchyma of palea. ×250.

at the apex of the grain. These hairs appear to have a kind of sheath around them, but this appearance seems to be due to the fact that there is necessarily a small space containing air between the otherwise closely adherent inner epidermis and epicarp, in the region of the hair. [Figs. 7 and 8.]

The hairs are 60–100 μ in length.

ii. *Epicarp and mesocarp.* These layers are not at all easy to see in preparations. The walls are not

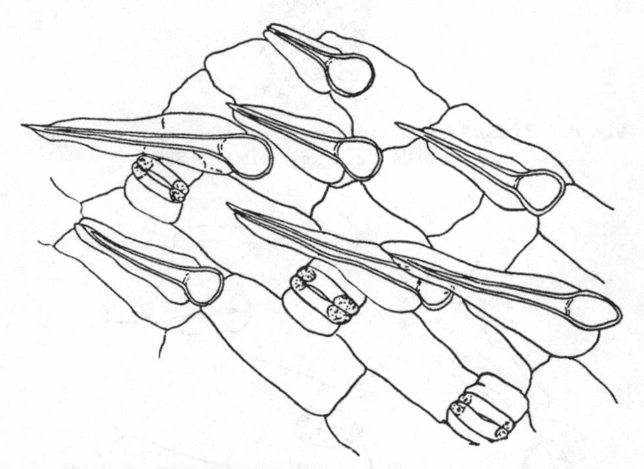

Fig. 8. *Barley.* Another view of inner epidermis of palea with hairs and stomata. The sheath-like appearance around the hairs is probably due to a cavity between the inner epidermis of the palea and the epicarp of the fruit. In some cases it contains air. ×250.

beaded and the epicarp possesses short, thick walled, awl-shaped hairs at the apex of the grain which are quite distinct in appearance from those mentioned above, being up to 250 μ in length. [Fig. 9.]

iii. *Cross cells.* There is a double layer of these

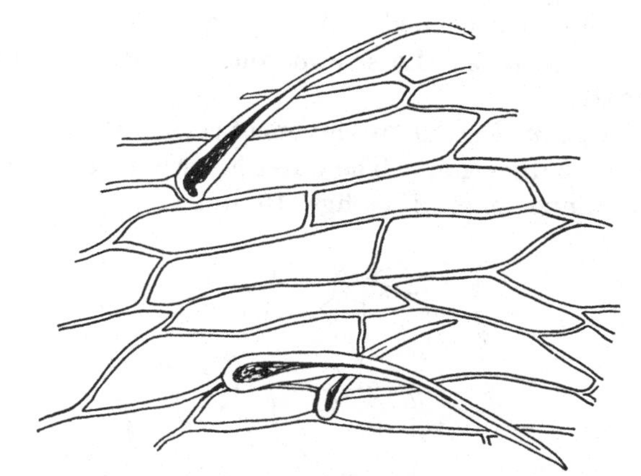

Fig. 9. *Barley.* Epicarp (surface view) with hairs. ×250.

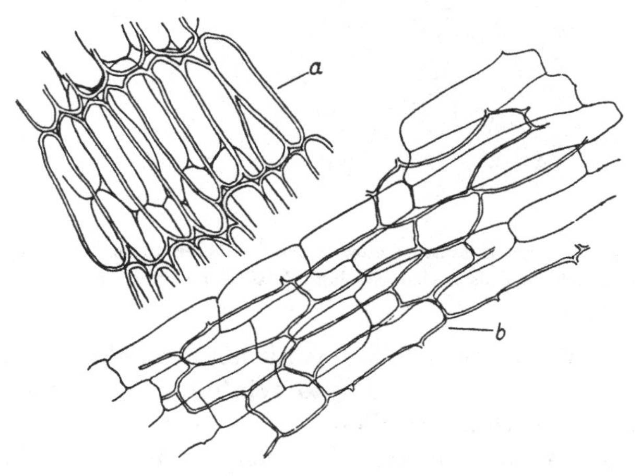

Fig. 10. *Barley.* *a*, cross cells. *b*, spermoderm (two layers). Surface view. ×250.

cells varying from 40 μ to 60 μ in length and possessing rather thin, non-beaded walls. [Fig. 10.]

iv. *Tube cells.* Less numerous and distinct than in wheat.

v. *Spermoderm* of two layers of cells both elongated in the same direction. The outer layer is more delicate than the inner one. [See figs. 10 and 11.]

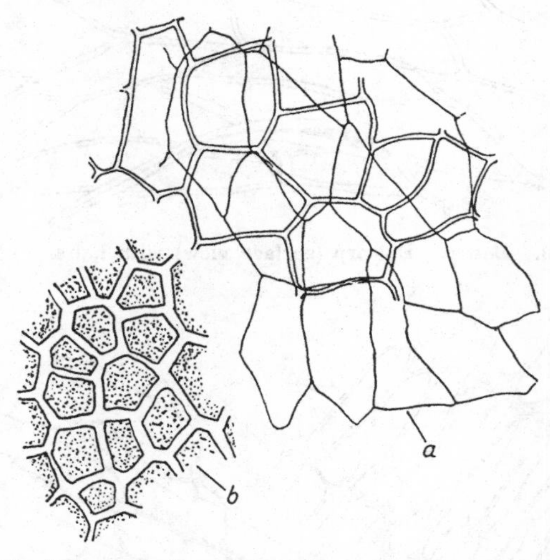

Fig 11 *Barley.* *a*, another view of the two layers of the spermoderm in which the cells are less regular. *b*, aleurone layer. Surface view. ×250.

vi. *Perisperm.* This layer is not evident in surface view but is seen, according to Winton, in section after treatment with dilute alkali.

vii. *Aleurone layer.* This consists of two or three layers of cells with darkly granular contents. The cells are smaller than those of wheat and the walls swell greatly on boiling with alkali. [Fig. 11.]

The starch grains are spherical and in two sizes. The large ones are distinctly smaller than those of wheat, there being few well-developed grains over about 35 μ. [See fig. 54 at end.]

Distinguishing characteristics.

i. Presence of closely adherent paleae with characteristic spongy parenchyma and hairs.

ii. Epicarp unbeaded, and indistinct. Hairs much shorter than in other cereals.

iii. Cross cells in double layer with unbeaded walls.

iv. Both layers of spermoderm running in the same direction and not crossing one another as in wheat.

v. Aleurone cells in two or three layers.

Oats. (Avena sativa L.)

i. *Paleae.* These are present in cattle-foods, but are removed from the grain in human foods. They are easily distinguished from those of barley by their shape and thickness. The outer epidermal cells have very thick walls and narrow lumens and the spongy parenchyma is stellate.

ii. *Epicarp and mesocarp.* The outermost layer of the fruit consists of thin-walled cells, more or less distinctly beaded, and greatly elongated, except at the ends of the grain. Hairs from 1 to 2 mm. long are borne singly or in groups of three, four, or more, over the whole surface. The basal part of the hair is often considerably narrower than the middle. [For epicarp, see fig. 12.] The mesocarp consists of several layers of thin-walled elongated cells more or less disintegrated in the ripe grain.

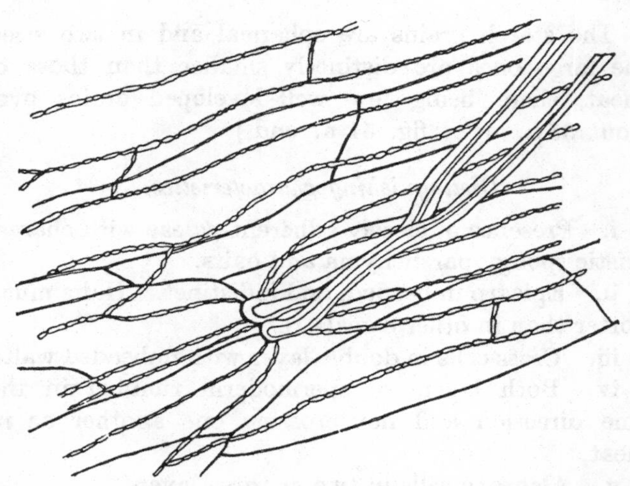

Fig. 12. *Oat.* Surface view of cells of epicarp and mesocarp. The cells of the epicarp have beaded walls and bear hairs. Surface view. ×250.

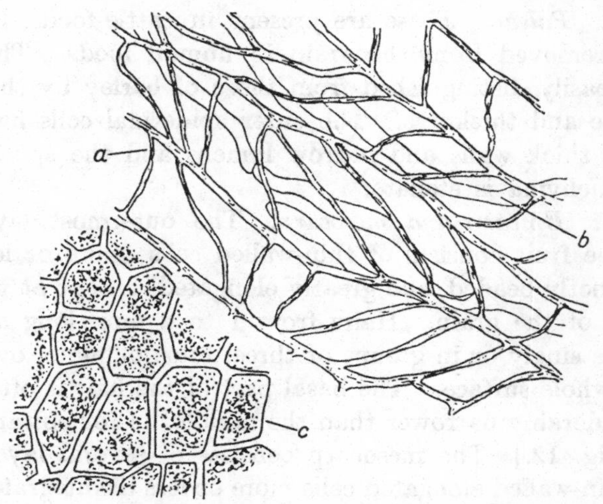

Fig. 13. *Oat.* *a,* cross cells. *b,* tube cells. *c,* aleurone cells in surface view. The long axes of the cross cells of one row are seen to make an obtuse angle with those of an adjacent row. ×250.

iii. *Cross cells.* These are somewhat irregularly set in a zig-zag manner in rows [see fig. 13]. They form a single layer and are not so distinct as in other cereals.

iv. *Tube cells.* Few and indistinct. [Fig. 13.]

v. *Spermoderm and perisperm.* Not evident in the ripe grain.

vi. *Aleurone cells.* Mostly in a single layer. [Fig. 13.]

Starch grains. The great majority of the grains are small (2–10 μ) and polygonal in shape and are aggregated into spherical or ellipsoidal masses which readily become broken up. A few small spindle-shaped grains are found among the others. [See fig. 54.]

Distinguishing features.

i. Paleae much thicker than those of barley, with thick walled outer epidermis and stellate parenchyma.

ii. Epicarp, with long hairs all over the surface.

iii. Cross cells indistinct and arranged in a peculiar zig-zag manner in rows.

iv. Starch grains small and polygonal, with spindle-shaped grains present.

Rice. (Oryza sativa L.)

i. *Paleae.* These are not present in rice intended for human consumption, but are generally found in cattle-foods which contain this cereal. They should not be present in too great quantities as they are harsh and brittle and possess little or no feeding value, although it is possible that they may act as stimulants to the walls of the intestine.

They are of a light yellow colour and have a ribbed

upper surface. The cells of the upper epidermis have very thick and deeply sinuous walls [see fig. 15]. They are not elongated longitudinally, but are often broader than they are long. Numerous stout, dagger-like hairs

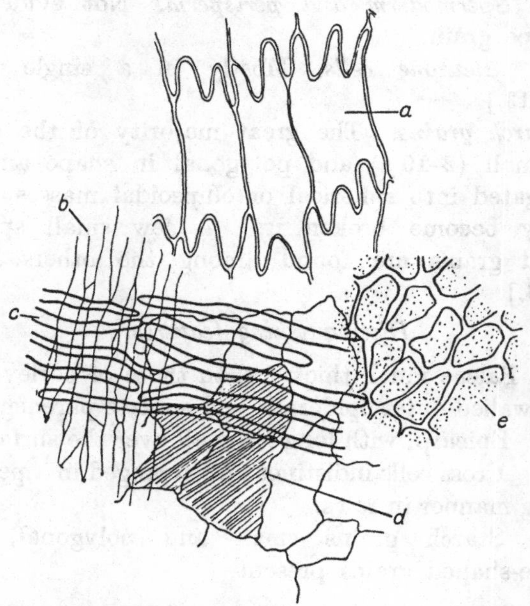

Fig. 14. *Rice*. Surface preparation of pericarp. *a*, epicarp with cells elongated transversely to the long axis of the grain. *b*, mesocarp and 'cross cells.' *c*, tube cells. *d*, spermoderm (pigmented). *e*, aleurone cells. The cells of the perisperm are not evident in this preparation. ×250.

which easily break off leaving round scars are borne among the epidermal cells.

The remaining layers of the paleae, viz., elongated fibres, spongy parenchyma and lower epidermis with stomata are masked, in ordinary preparations, by the upper epidermal cells to which the conspicuous and characteristic appearance of the paleae is entirely due.

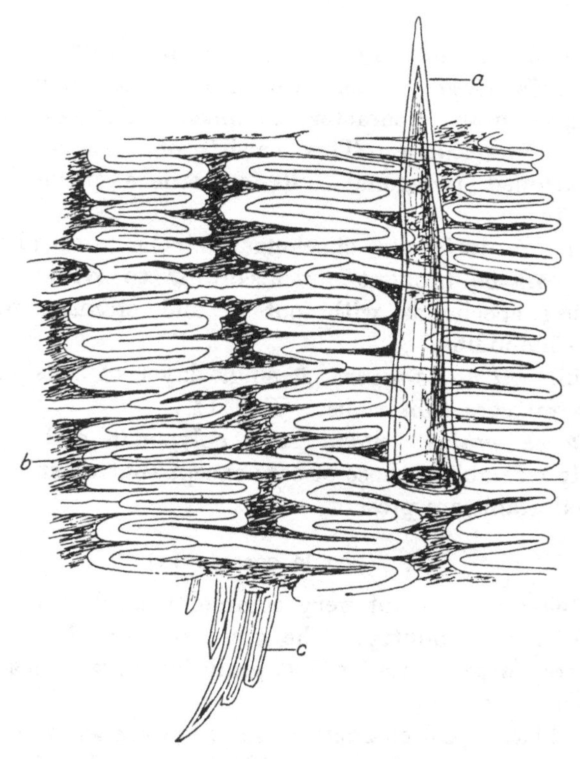

Fig. 15. *Rice palea.* Outer epidermis surface view. *a*, dagger cells. *b*, wavy cells. *c*, fibres. ×250.

ii. *Epicarp.* This differs from the epicarp of other cereals in having its cells transversely instead of longitudinally elongated. The cell-walls are non-beaded, and the short longitudinal ones are waved. [Fig. 14.]

iii. The cells of the *mesocarp* are transversely elongated (increasingly so towards the centre of the grain), and are indistinguishable from the cross cells. [Fig. 14.]

iv. *Cross cells.* These do not form a distinct layer.

v. *Tube cells.* These are distinct and numerous, giving the bran a characteristic appearance. [Fig. 14.]

vi. *Spermoderm.* This consists of a single layer of flattened polygonal cells sometimes pigmented. [Fig. 14.]

vii. A *perisperm* of elongated cells with pitted walls can be brought out according to Winton by treating specimens with potash and staining with chlor-zinc-iodine.

viii. *Aleurone layer.* A single layer of cells with comparatively thin walls. [Fig. 14.]

Starch grains. Small and polygonal and very closely resembling those of oats except that no spindle-shaped grains are present.

Rye. (Secale cereale L.)

This cereal is not very commonly used in cattle-foods in this country. The grain is very similar in appearance to a small wheat grain, and like wheat it threshes out clean.

In histological characters rye resembles wheat more closely than any other grain. It is distinguished however by the following characters:

1. The hairs on the epicarp are not quite so stout as in the case of wheat. They are also less sharply pointed. [Figs. 16 and 17.]

2. The walls of the mesocarp cells are not beaded. [Fig. 18.]

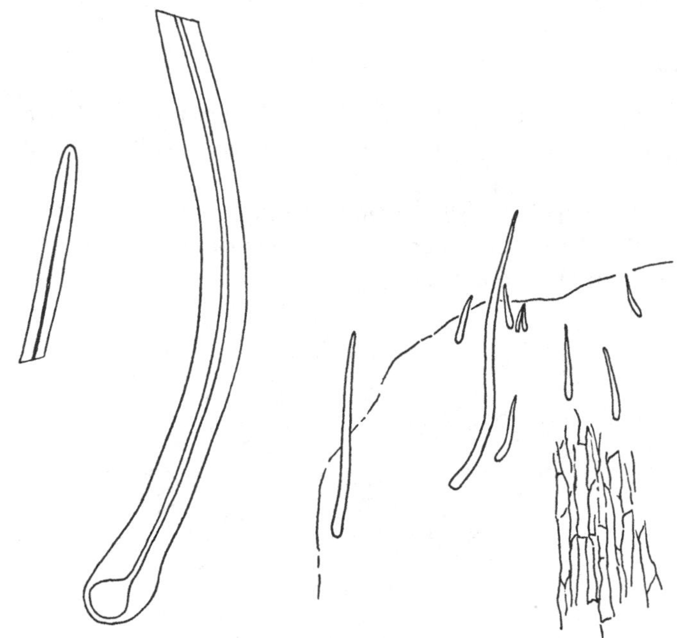

Fig. 16. Tip and base of hair from epicarp of rye. ×250.

Fig. 17. Hairs on epicarp of rye ×50.

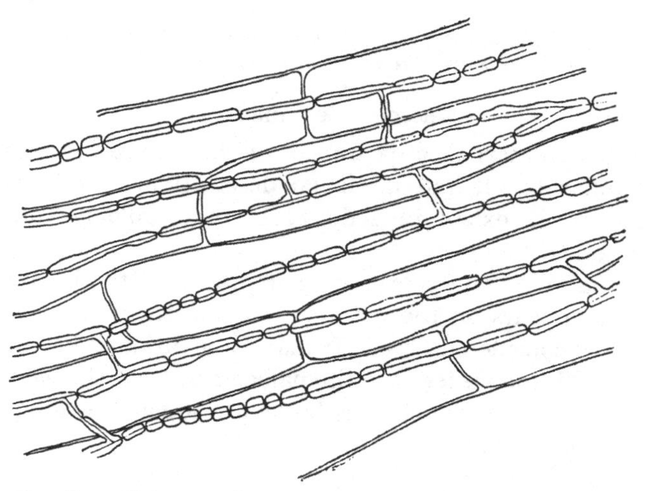

Fig. 18. *Rye.* Epicarp and mesocarp. The mesocarp cells are non-beaded. Surface view. ×250.

3. The walls of the cross cells are not beaded at the ends of the cells, but are slightly swollen in this region. [Fig. 19.]

4. The *tube cells* are not so numerous as in wheat.

5. The spermoderm and perisperm are hardly seen at all in the ripe grain.

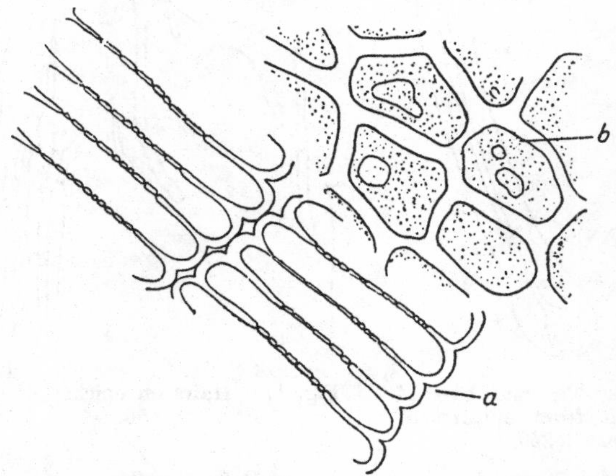

Fig. 19. *Rye.* *a,* cross ce'ls. *b,* aleurone cells. Surface view. The ends of the cross cells are swollen and non-beaded. ×250.

Maize. (Zea mays L.)

The external appearance of the maize grain is familiar to all and needs no description. As is well known, it exists in red, white, and golden yellow varieties.

Paleae. The paleae are not present with the grain as it appears on the market; consequently they need not be considered here. It may be mentioned however that Winton states that the paleae and even the woody cobs on which the grain is borne are sometimes used as food adulterants.

Pericarp. The pericarp is horny and transparent, and for examination the grain should be soaked in water until it is soft enough to be dealt with. Boiling with potash is apt to swell the tissues and make them somewhat indistinct, hence in examining maize this process may be omitted.

Fig. 20 shows a transverse section through the pericarp. The cells of the *epicarp* and *mesocarp* are

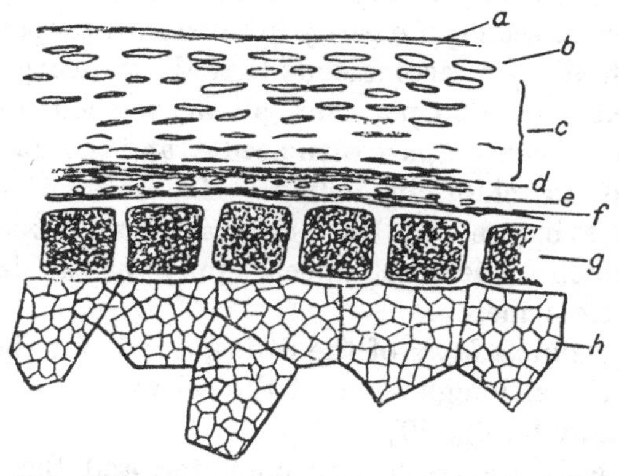

Fig. 20. *Maize.* Transverse section of pericarp. *a*, cuticle. *b*, epicarp. *c*, mesocarp. *d*, parenchyma. *e*, tube cells. *f*, spermoderm and perisperm. *g*, aleurone layer. *h*, outer starch cells. ×333.

seen to be thick walled with narrow lumens, the former being covered on the outside by a cuticle.

The cross cells are represented by a spongy parenchyma (*d*), the tube cells are cut across at *e* and an indistinct spermoderm and perisperm are shown at *f* between the tube cells and the aleurone layer (*g*).

Surface preparations may be made by peeling off thin layers from the soaked grains, or by cutting surface sections from flat portions of the grain with a sharp razor.

The cells of the epicarp and mesocarp are longitudinally elongated and have thick, beaded walls [see fig. 21]. The *spongy parenchyma* is shown at *b* in the same figure and at *a* in fig. 22. It will be seen that the axis of elongation of the cells of this layer is in a transverse direction, especially in fig. 22 which represents a few of the innermost cells.

The *tube cells* [fig. 22] are numerous and very long. They can be seen quite easily in unstained preparations although Winton recommends that the pericarp should be boiled with 1·25 per cent. alkali, washed in dilute acetic acid, picked apart with needles and the fragments mounted in chlor-zinc-iodine. He states that this treatment brings out the *spermoderm* and *perisperm* also, which otherwise appear only as a delicate, structureless membrane.

The *aleurone layer* of the endosperm is for the most part single, although some of the cells may be divided tangentially [*g*, fig. 20].

The cells are 30–40 μ in diameter, and the double walls 6–9 μ thick.

Starch. The cells containing the starch increase in size towards the centre of the grain as do the starch grains which they contain.

The grains are polygonal or rounded, and from 10–30 μ or more in diameter. Nearly all of them have a central hilum with radiating clefts. [See fig. 54.]

Millet. (Panicum miliaceum.)

Millet is cultivated for grain in China, India, Japan, and to some extent in Europe and America. In this

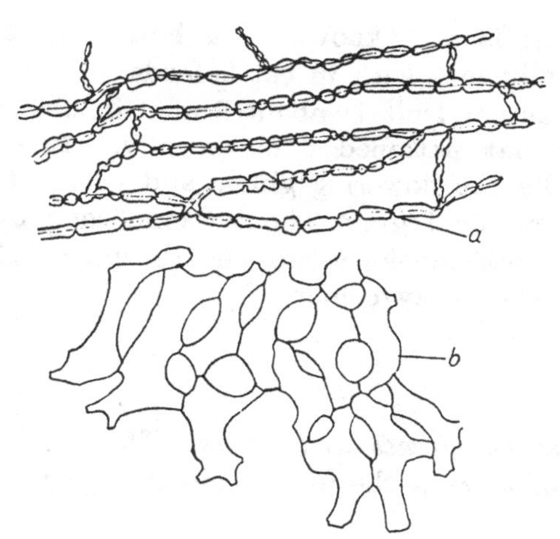

Fig. 21. *Maize.* Surface preparation. *a*, mesocarp cells. *b*, parenchyma. ×250.

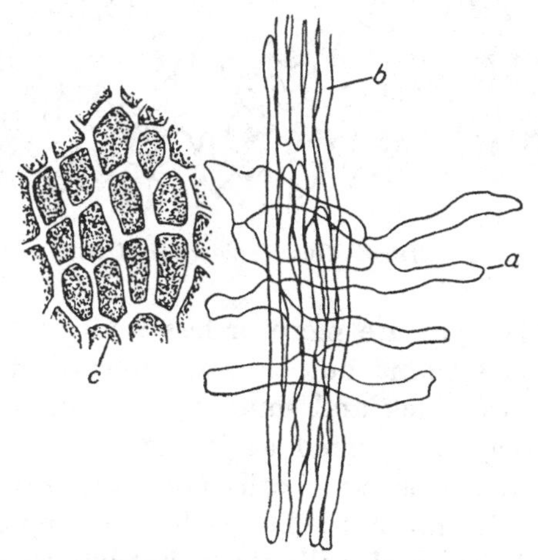

Fig. 22. *Maize.* Surface preparation. *a*, parenchyma. *b*, tube cells. *c*, aleurone cells. ×250.

country it is best known as a bird seed, but it is occasionally met with in cattle-foods.

The actual fruit is about 2 mm. in diameter and is somewhat flattened dorsiventrally. It is tightly clasped by the flowering glume and palet, the whole forming an oval grain about 3 mm. long by 2 mm. broad. Both envelopes are smooth and polished and of a uniform straw-colour.

Histology.

Flowering glumes and palets. The most striking feature of these is the upper epidermis [fig. 23]. The

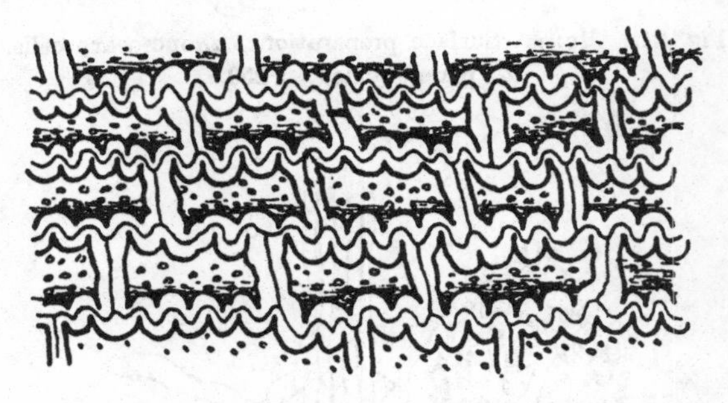

Fig. 23. *Millet.* Outer epidermis of palet. ×300.

cells of this layer are more or less isodiametric in the centre of the glume but elongate towards the edges. The walls are thick and sinuous; the cell-lumens are not so nearly obliterated as is the case in rice, and neither round cells, twin cells nor hairs are present.

Epicarp [*a*, fig. 24]. The cells of the epicarp have sinuous side and end walls somewhat like those of rice. They differ from the latter in being less sharply sinuous;

the axis of elongation also is longitudinal with respect to the grain instead of being transverse.

Cross cells and tube cells [*b* and *c*, fig. 24]. These are both very much elongated. They show up distinctly in some preparations although it is difficult to make out their actual length.

Perisperm. Thin-walled elongated cells probably

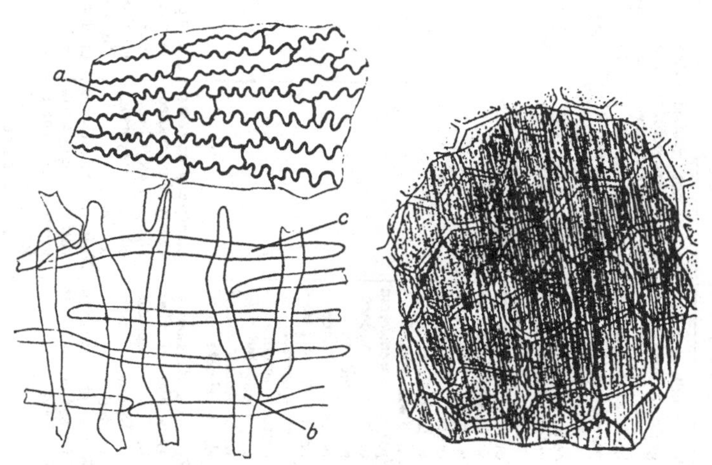

Fig. 24. *Millet.* Pericarp. *a*, epicarp. *b*, cross cells. *c*, tube cells. ×250.

Fig. 25. *Millet.* Aleurone cells and perisperm. ×250

belonging to the perisperm are shown overlying the aleurone layer in fig. 25.

The *aleurone layer* is of the usual type.

The *starch grains* are polygonal in outline, and are about the same size as those of rice.

Millet is recognised by means of the flowering glumes and palets, which are distinguished from those of barley, wheat, oats, darnel, chess, rice, and maize, by the absence of round cells, twin cells, and hairs;

Table for Recognition of the commonest Cereals.

Wheat	Rye	Barley	Oats	Rice	Maize
Paleae not present in foods.	*Paleae* not present in foods.	*Paleae* adherent to grain. Ribbed, light yellow *Outer epidermis* of paleae consists of (i) Wavy elongated cells with moderately thick walls. (ii) Round cells with outer tangential walls protruding as a blunt hair and inner tangential walls pitted. (iii) Twin cells. *Spongy parenchyma* of irregular square-shaped cells in surface view, with many inter-cellular spaces. *Inner epidermis* of delicate longitudinally elongated cells with stomata and short delicate hairs.	*Paleae* not ribbed. May be dark-coloured or light yellow. *Outer epidermis of* elongated wavy cells with very thick walls. Many layers of *fibrous cells* present. *Spongy parenchyma* stellate	*Paleae* ribbed. *Outer epidermis of* very thick walled wavy cells hardly longer than broad (see fig. 15). *Characteristic dagger-like* hairs present.	*Paleae* not present in foods, except as impurities.
Epicarp and mesocarp. Thick walled cells, elongated in longitudinal direction in middle belt of grain, more isodiametric at ends. *Walls* beaded all round in *both* layers. *Hairs* borne only at distal end; thick walled, unicellular, pointed	*Epicarp and mesocarp* Cells thinner walled than in wheat. *Epicarp only* beaded. *Hairs* somewhat delicate. Ends rounded.	*Epicarp and mesocarp.* Less distinct than in other cereals. Cell-walls, non-beaded. *Hairs* shorter and fewer than in wheat and rye.	*Epicarp and mesocarp.* Fairly thin walled. *Epicarp only* beaded. Inner layers indistinct. *Hairs* borne all over surface singly or in groups of three, four, or more. Much longer than in other cereals.	*Epicarp.* Of TRANSVERSELY elongated cells with wavy longitudinal walls. Mesocarp passes gradually into cross cells.	*Epicarp and mesocarp.* No hairs on epicarp. Cells thick walled, beaded; outer cuticle thin. Seven or eight layers of mesocarp cells longitudinally elongated; walls beaded.

Cross cells. Single layer, fairly regular, thick walled, beaded all the way round. No intercellular spaces.	*Cross cells.* Single layer, regularly arranged. Not so thick walled as in wheat, and beading not carried round the ends of the cells. These ends swell in potash and show well-marked intercellular spaces.	*Cross cells.* Double layer. Cells smaller than in wheat and rye. Walls relatively thin and not beaded.	*Cross cells.* Single layer of large cells with non-beaded walls. Arranged in a zig-zag manner in rows. [See fig. 13.]	*Cross cells.* Not differentiated from the inner layers of the mesocarp.	*Cross cells.* Represented by a spongy parenchyma.
Tube cells. Conspicuous and very numerous in some parts of the grain.	*Tube cells.* Not numerous.	*Tube cells.* Not conspicuous and not numerous.	*Tube cells.* Very few and indistinct.	*Tube cells.* Very numerous and distinct forming the most characteristic layer in the bran.	*Tube cells.* Present. Seen when epicarp and mesocarp are removed
Spermoderm of two layers of thin walled, elongated cells, crossing one another.	*Spermoderm* very indistinct.	*Spermoderm.* Upper layer of thin walled cells, indistinct, lower layer thicker walled and fairly distinct. The cells in both these layers are elongated longitudinally.	*Spermoderm* not seen in ripe grain.	*Spermoderm* distinct and sometimes pigmented polygonal cells.	*Spermoderm* seen best in section. Two layers.
Perisperm not usually seen in surface preparations.	*Perisperm* not visible in surface preparations.	*Perisperm* not visible in surface preparations.	*Perisperm* not seen in surface preparations.	*Perisperm* not easily seen in surface preparations.	*Perisperm* seen in surface view, when outer layers are removed.

It is not easy to distinguish between the *aleurone layers* of the different cereals in surface preparations.

from those of German millet by the absence of wrinkles on the outer epidermis and from those of green foxtail by the absence of both wrinkles and patches of brown tissue[1].

Cereals as Cattle-foods.

Wheat.

(*a*) *Whole wheat* crushed is sometimes fed to horses, cattle, and pigs, when wheat is cheap. It has also been found crushed in a mixed cattle cake and in various lamb foods.

(*b*) *Wheat meal*, consisting of whole wheat ground up, is used sometimes as a pig food.

(*c*) *Wheat bran* consists of the outer coats of the wheat grain with a certain amount of the starchy endosperm adhering to them. It is largely used as a pig food, but is also found in cattle cakes, calf meals, pig meals and lamb foods.

(*d*) *Wheat middlings or pollards* consist of the finer bran particles together with a varying quantity of starchy endosperm. It is used chiefly for pigs.

(*e*) *Wheat germ* consists of the embryo at the base of the grain together with the overlying bran coats. It is used as food both for cattle and human beings.

Barley.

(*a*) *Barley meal*, consisting of the coarser types of barley ground up, is the principal pig food in this country. It contains the paleae as well as the grain.

(*b*) *Brewer's grains* consist of the remains of the barley after it has passed through the process of

[1] Winton, p 117.

fermentation. They are frequently used as a cattle food and as constituents of patent food mixtures.

(c) *Malt culms.* In the process of malting, barley is allowed to send out rootlets from ¼ to ½ an inch long, and then the growth is stopped. These short rootlets get rubbed off in the treatment which the malt undergoes and are collected and sold as malt culms. Their feeding value is not very great, but they have been found in feeding cakes and in other patent food mixtures.

Rice.

Coarsely ground rice is the commonest starchy ingredient in mixed cattle cakes, the presence of paleae, bran, and small polygonal starch grains affording a ready means of detecting it.

Rice bran and middlings are also largely used in these preparations.

Oats.

Whole or crushed oats are an extremely important food for horses, cattle and sheep in this country. For human consumption the paleae are removed.

Maize.

Whole or crushed maize is used for pigs, cattle and poultry.

Maize germ. The germ of maize contains oil but no starch. It is removed from the rest of the grain on a commercial scale, and after the oil has been expressed it appears on the market as maize germ meal and maize germ cake, both of which contain some residual oil. The meal is also introduced into mixed feeding cakes and calf meals.

According to Winton, *maize cobs* are often ground

and employed as an adulterant of other foods in the United States. Maize bran has also been used to adulterate wheat bran.

Maize flour or meal is sometimes used instead of rice meal in mixed cattle cakes.

CHAPTER III

LEGUMINOUS PLANTS OR PULSES

The leguminous seeds commonly used as cattle foods are Soya beans, Ground nuts (also called Earth nuts, Pea nuts and Monkey nuts), Carob beans (also known as Locust beans), Horse beans, and Peas, whilst Fenugreek is used as a flavouring material.

A typical leguminous seéd-coat in transverse section is made up of the following layers of cells:

(*a*) An outermost layer of radially elongated, regularly arranged cells, known as *Palisade cells.* The lumens of these are as a rule filled in towards the outside by the thickening of the cell walls; towards the inside however the walls are thinner and pigmented contents may be present. A refractile band known as the 'light line' is seen at a little distance from the outside.

(*b*) *The sub-epidermal or hypodermal layer* (also known as *column cells*) is generally highly characteristic in these seeds. In peas and horse beans the cells appear dumb-bell shaped in section [see figs. 26 and 33]. In soya they are hour-glass shaped [see fig. 27]. Those of fenugreek have very characteristic thickenings on their walls [see figs. 33 and 34] while those of *Phaseolus vulgaris* (the common French bean) are polygonal in

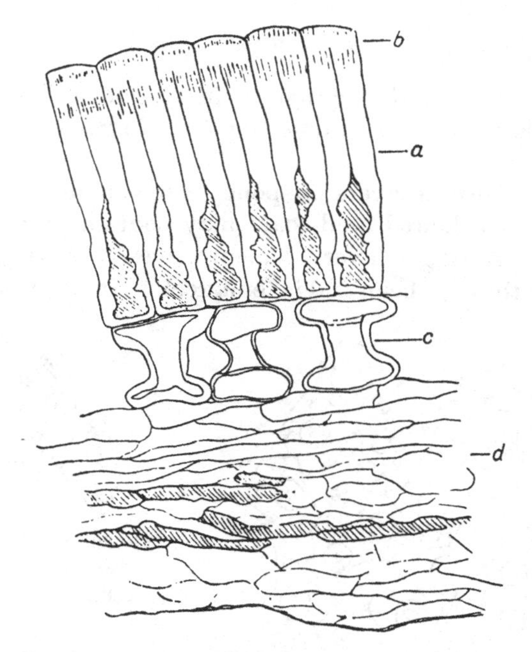

Fig. 26. *Horse bean* (Faba vulgaris). Transverse section of seed coat. *a*, palisade cells with light line indicated at *b*. *c*, column cells. *d*, parenchyma. Shading represents brown pigment. ×250.

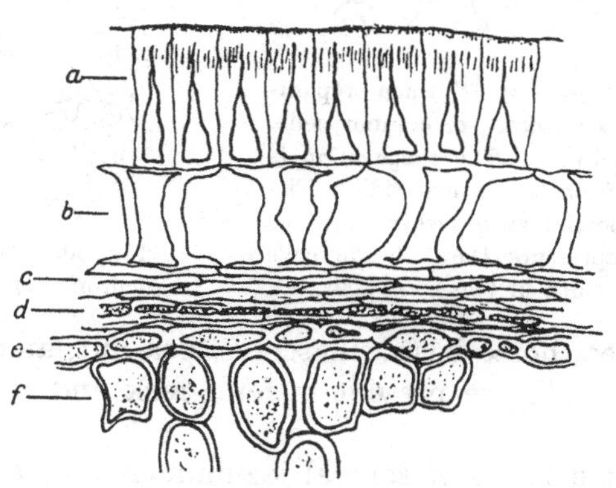

Fig. 27. *Soya bean* (Soja hispida Moench). Transverse section of seed coat. *a*, palisade cells. *b*, column cells. *c*, parenchyma. *d*, aleurone cells. *e*, epidermis of cotyledon. *f*, cotyledonary tissue. ×250.

surface view and rectangular in section and contain crystals of calcium oxalate [fig. 31] of striking appearance.

(*c*) Several layers of spongy parenchyma, some of which in coloured seed-coats may contain pigment.

(*d*) A single layer of aleurone cells is very well seen in the soya bean and in fenugreek. It is not present

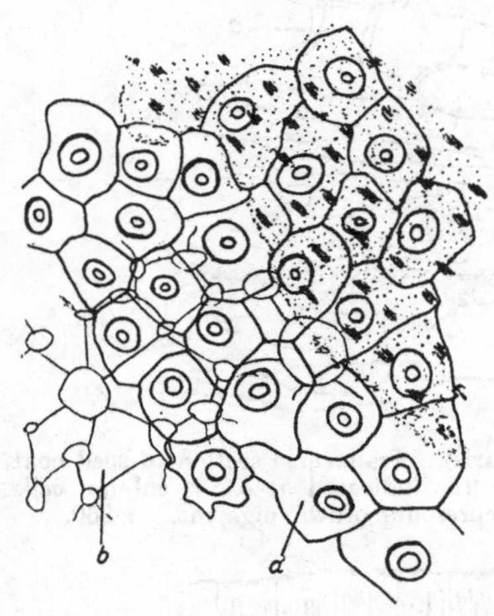

Fig. 28. *Soya bean.* Surface preparation of seed-coat. *a*, column cells. *b*, parenchyma. The preparation is viewed from the inner side. The shaded portion with darker markings on it represents the palisade cells somewhat out of focus. ×300.

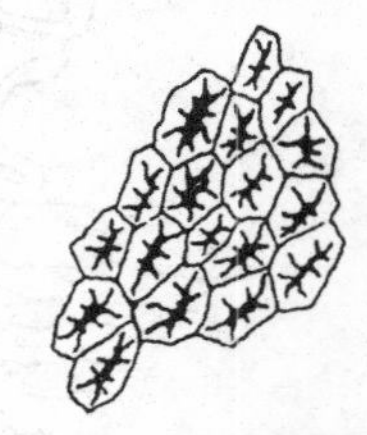

Fig. 29. *Soya bean.* Surface view of palisade cells. ×300.

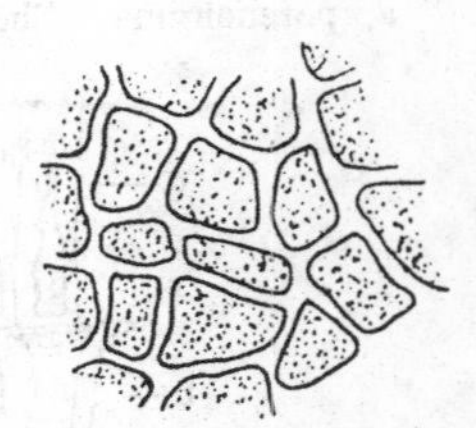

Fig. 30. *Soya bean.* Aleurone cells. ×300.

in the pea, horse bean, phaseolus, pea nut, and locust. *Mucilage cells* are present in fenugreek and the carob bean.

Surface views of several leguminous seed-coats are shown in figs. 28, 29, 30, 31, 32, 33, 35. It will be seen that the lumens of the *palisade cells* are branched. If however the outer wall of these cells is in focus a reticulated appearance shown in fig. 31 is presented.

The narrow middle part of the dumb-bell and hourglass shaped cells stands out in surface view as a ring,

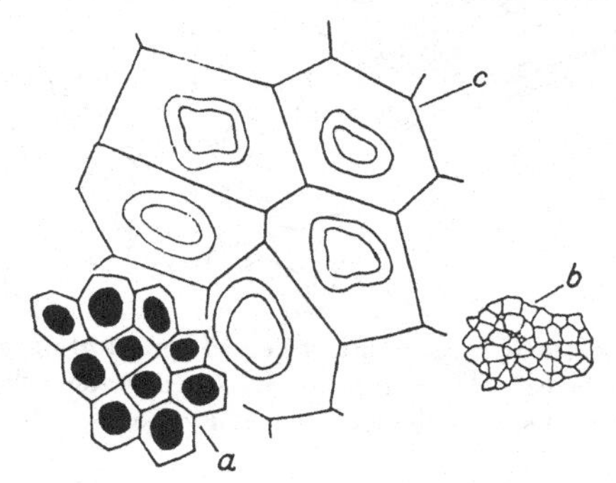

Fig. 31. *Horse bean.* Surface preparation. *a*, palisade cells (lower portion in focus). *b*, palisade cells (outer surface in focus). *c*, column cells. ×250.

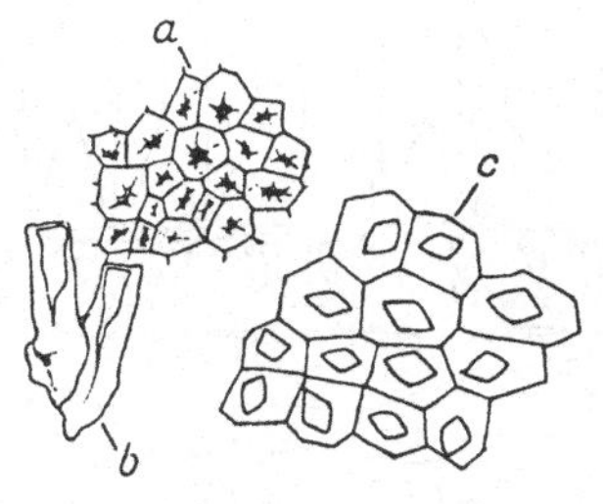

Fig. 32. *Common bean* (Phaseolus vulgaris). *a*, palisade (surface view). *b*, palisade (two isolated cells. side view). *c*, hypodermal cells with crystals. ×250.

which is especially prominent in the soya bean [see fig. 28]. The crystal-containing hypodermal cells of phaseolus are shown in fig. 32.

The hypodermal cells of fenugreek [fig. 35] are

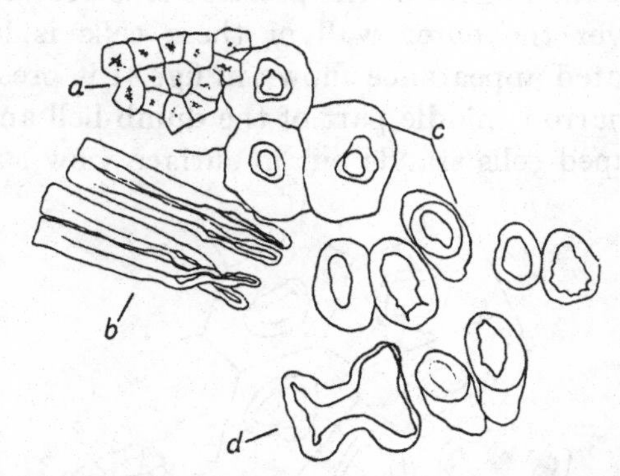

Fig. 33. *Pea* (Pisum sativum). Teased preparation of seed-coat. *a*, palisade cells in surface view. *b*, palisade cells, side view. *c*, column cells, surface view. *d*, column cells, side view. ×250.

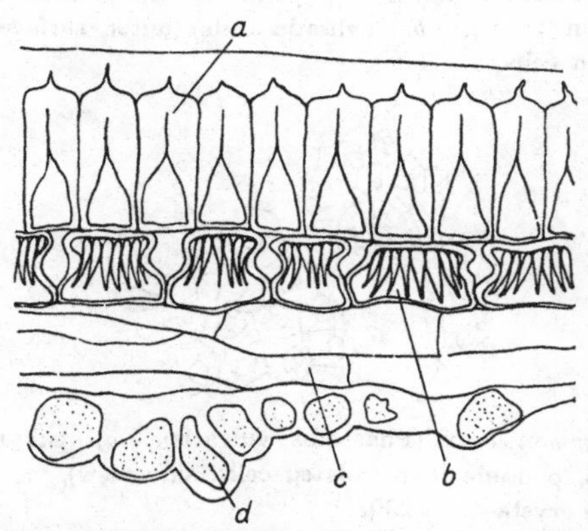

Fig. 34. *Fenugreek* (Trigonella Foenum-Graecum). Transverse section of seed-coat. *a*, palisade with cuticle. *b*, column cells with curiously thickened walls. *c*, parenchyma. *d*, aleurone layer. ×250.

strikingly beautiful in surface view. Its palisade cells are peculiar in being pointed and covered by a fairly thick cuticle (fig. 34).

The carob bean and ground nut require separate consideration from the other leguminous seeds because the

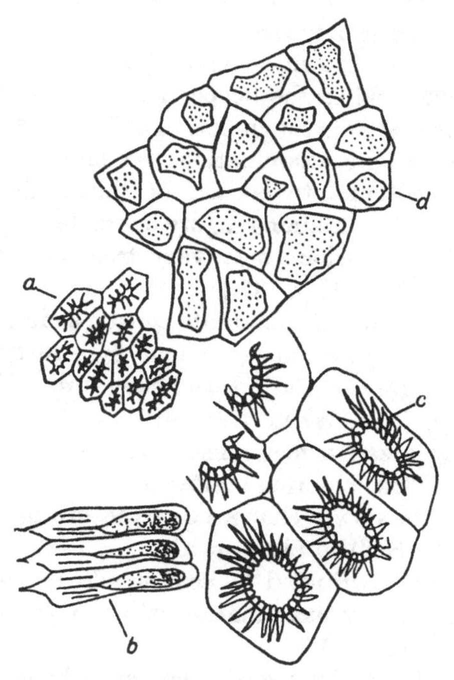

Fig. 35. Surface preparation of seed coat of *Fenugreek*. *a*, palisade cells, surface view. *b*, palisade cells, side view *c*, hypodermal or column cells, surface view. *d*, aleurone layer—walls swollen with potash. ×333.

fruit walls as well as the seeds themselves are present in cattle-foods, and also because the seed-coats do not altogether conform to the general type.

The carob bean or locust bean. (Ceratonia Siliqua L.)

The husks or shells of the carob bean have a pleasant, sweet taste and are very commonly found in mixed cakes, lamb foods, etc. When present in large fragments they are easily recognised in the original material by their appearance and taste. They have a dark, reddish brown—almost black—leathery and polished outer skin; the deeper layers are dull brown and the inner skin is tough and light brown in colour.

In microscopic examination the middle layers of the husk are found to consist of thin-walled parenchymatous cells containing highly characteristic 'brown wrinkled bodies' which may vary from $30\,\mu$ to $230\,\mu \times 25\,\mu$ to $100\,\mu$. [Fig. 37.]

In the presence of cold dilute caustic soda or caustic potash these bodies change colour first to green and then to blue grey which in turn changes to violet when heat is applied. When they are cautiously heated with strong alkali a magnificent deep blue is at once obtained. This colouring matter is insoluble in alcohol and ether, but slowly changes on exposure to air (more quickly with HCl) to red brown[1].

The 'brown wrinkled bodies' should be sought for in the preliminary examination of the powdered food material. [See p. 7.]

The seeds of the locust bean are about ¼″ long, flattened from side to side, and of a dark reddish brown colour. They possess a fairly high polish.

They are extremely hard, and owing to the difficulty of grinding them they are often present in rather large fragments. In section the seed coat is seen to be

[1] Winton, *Microscopy of Vegetable Foods*, p. 276.

extremely thick, the palisade cells alone being 150 μ to 250 μ long, the hour-glass cells 20–40 μ, while the parenchyma is much thicker and more compact than it is in such seeds as the pea and bean.

The endosperm is a dense horny layer of a greenish grey colour and encloses a flattened, yellow embryo.

When the seeds are boiled with water or potash the palisade and column cells peel off, leaving only the parenchyma adhering to the endosperm; if boiling is prolonged this may come off also.

The endosperm swells forming yellowish, translucent masses easily recognised in a mixture.

Ground nut.

[Synonyms: pea nut, earth nut, monkey nut.] The ground nut is the fruit of a leguminous plant *Arachis hypogea*, and is so called from the fact that the pods, after fertilization has taken place, bury themselves in the ground to ripen, root hairs even being developed on the epidermis.

Cakes made from the ground nut are in two main classes:

(*a*) *Decorticated varieties* in which the pod is more or less completely removed. (Search with the microscope will always reveal some fragments of it.)

(*b*) *Undecorticated* varieties which will contain the pods and probably fragments of the stems and leaves of the plants as well.

A common impurity in these cakes is sand on account of the fact that it is difficult to remove all the earthy matter attached to the pods.

The most characteristic elements in the ground nut for identification purposes are as follows:

A. In the pods.

(1) Crossing layers of fibres, which when macerated and teased out are seen to be in many cases curiously

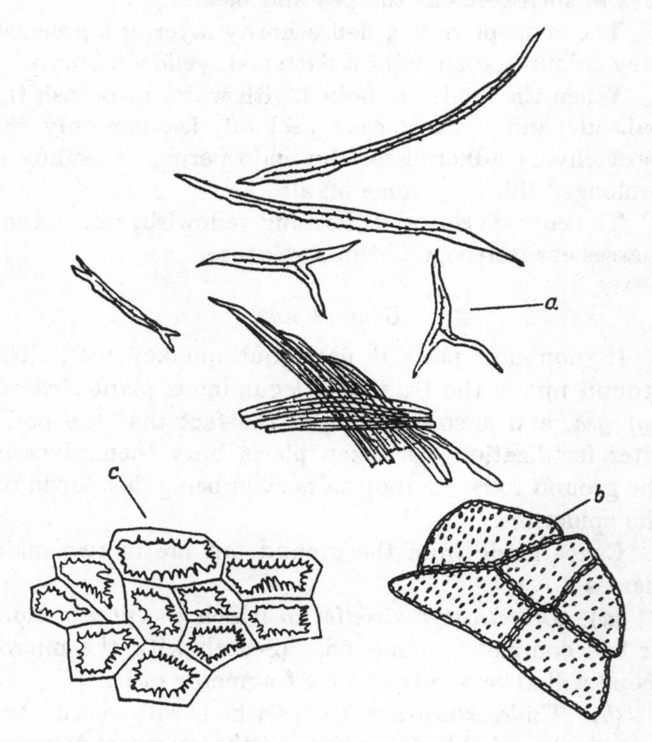

Fig. 36. *Ground nut* (Arachis hypogea). *a*, fibres of various shapes from the fruit wall. *b*, hypodermal cells from fruit wall. *c*, epidermal cells of seed-coat. $a \times 66$. *b* and $c \times 333$.

shaped and branched. Some of them have saw-like edges marking the points of contact with the layer beneath. Various forms of fibres are shown in fig. 36.

(2) Curious pitted hypodermal cells (see fig. 36). The root-hairs cannot be seen in the pea nuts as sold, the epidermis being mostly rubbed off. The pods can be studied either by boiling small fragments in potash and teasing with needles, or by teasing after macerating in Schulze's solution.

B. In the brown seed coats.

The colourless epidermal cells with walls unevenly thickened and pitted towards the outer surface, form by far the most characteristic layer of the seed-coat, and although superficially they do not appear to do so, they really correspond with the palisade cells of other legumes. They are shown in surface view in fig. 36.

The underlying layers have no striking characters and no outstanding sub-epidermal layer exists.

The cotyledonary tissue contains starch, oil, and aleurone grains, the starch grains being small and spherical with a central refractile dot representing the hilum.

Starch grains of pulses.

The pea, horse bean, and French bean, possess typical leguminous starch, viz. oval or ellipsoidal grains with a branched crack representing the hilum [see fig. 54]. The starch grains of the ground nut have already been described.

The soya bean, carob bean, and fenugreek contain no starch.

Pulses as cattle-foods.

Horse beans form one of the staple foods for horses in this country and are fed split or whole. Coarsely ground beans may also be fed to cattle. Peas are not

Table comparing Leguminous Seeds.

	Horse bean	Phaseolus vulgaris	Pea	Soya	Locust	Ground nut	Fenugreek
Palisade	120 μ–130 μ high; 16–17 μ broad	50–60 μ high	60–100 μ high	50 μ high; 6–15 μ wide	160 to more than 200 μ high; 30–50 μ wide	Represented by characteristic epidermal cells of seed coat	60–70 μ high; 8–20 μ wide. Outer ends pointed
Sub-epidermal layer	Cells dumb-bell shaped. Up to 50 μ high	Cells prismatic. Each containing a crystal	Cells dumb-bell shaped about 30 μ high	Cells narrowed in the middle but not dumb-bell shaped. Very conspicuous in surface preparations. 40 μ high × 18 μ wide	Cells swell greatly when boiled. Dumb-bell shaped	No characteristic layer	Cells with characteristically thickened walls
Parenchyma	Spongy	Spongy	Spongy	Spongy	Thick layer of compact cells	Spongy	Spongy
Endosperm	Not present	Not present	Not present	Aleurone layer distinct in surface view	A thick horny layer. Walls of cells curiously thickened	Not present	Aleurone and mucilage cells present
Starch	Present. Grains ellipsoidal. Up to 70 μ in length	Present. Grains ellipsoidal. Up to 60 μ in length	Present. Ellipsoidal. Up to 60 μ long in some varieties	Absent	Absent	Grains small and spherical. Up to 10 μ in diameter	Absent

so commonly fed to animals, being usually somewhat expensive, and neither peas nor beans are used to any extent in cakes, although peas and also *Phaseolus vulgaris* have been found in them.

Soya beans are present in almost all mixed cakes

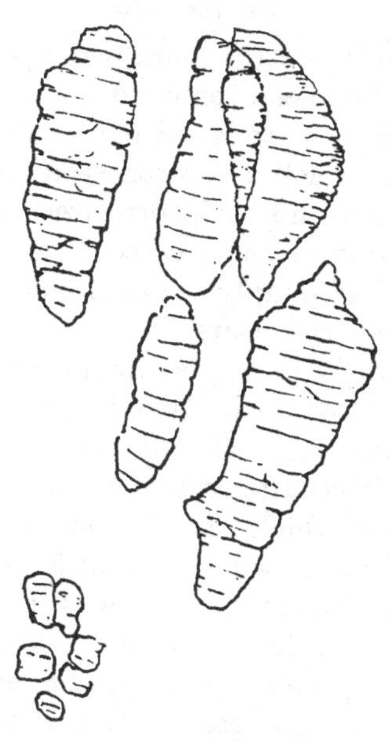

Fig. 37. *Carob bean.* Brown wrinkled bodies of fruit wall. ×100.

and ground nuts and carob beans also figure in them very commonly, whilst fenugreek is probably the commonest flavouring material. The latter may be present in such quantities in a cake as to be of some feeding value.

CHAPTER IV

OIL SEEDS

Cotton seed.

The seeds of the cotton plant (*Gossypium herbaceum*), after the fibre has been removed and the oil expressed by hydraulic means, are used for making cattle cakes. Cakes consisting solely of cotton seed are of two kinds, viz., decorticated and undecorticated. In the former a large proportion of the seed-coats is removed, in the latter the seed is simply crushed and no attempt is made to remove the coats.

Cotton seed is also present in nearly all mixed feeding cakes as an inspection of the examples given on the first two pages of this book will show.

The actual seed after ginning is about ¼″ in length, and possesses a thick, hard and somewhat brittle seed-coat of dull brown or blackish brown colour, to which a certain amount of fibre still adheres. Practically the whole of the seed-coat is filled by the embryo itself, the endosperm and perisperm existing only as a very thin skin. In cross section the cotyledons are seen to be much folded and are dotted with numerous dark coloured resin cavities. If a section is placed in water this resin flows out as a yellowish emulsion, the particles of which can be seen under the microscope to be in rapid motion.

Strong sulphuric acid, as was observed by Hanausek dissolves the secretion forming a blood-red solution. Alkalies colour it greenish brown but do not dissolve it. This reaction with sulphuric acid is one of the tests for cotton seed in a mixture[1].

The characters of the seed coats, however, serve as the chief means of identifying cotton seed. Fragments of them can generally be recognised quite easily by a surface examination of potash preparations. If there is any doubt, the fragments are usually large enough to be sectioned by hand, when the following structures are revealed [see fig. 38]:

(*a*) An *epidermis* of large, thick walled, and somewhat irregular cells with dark coloured contents.

(*b*) *Fibres* interspersed among, and occupying the position of, epidermal cells.

(*c*) *Parenchymatous cells* somewhat flattened and disorganised in ordinary hand sections of the ripe seed-coats, and possessing brown contents.

(*d*) *Vascular bundles.* These are surrounded by the above mentioned parenchymatous cells and correspond in position to very slight elevations on the surface of the seed.

(*e*) A layer of *thick walled colourless cells* presenting a characteristic appearance in surface view.

(*f*) Two layers of *palisade cells*, the upper ones being about 65 μ and the lower ones 130 μ in length. The lumens in both layers are largely obliterated except at the bases of the upper cells where they widen out and are rendered conspicuous by their brown contents. The general colour of the palisade cells in potash preparations is a pale yellow.

(*g*) *Parenchyma*, in several layers, with brown contents, is present within the palisade cells.

[1] For further chemical tests see *Agricultural Chemistry* by T. B. Wood, p. 52.

A perisperm, and an endosperm of typical aleurone cells, are also present, but are not seen in a section of

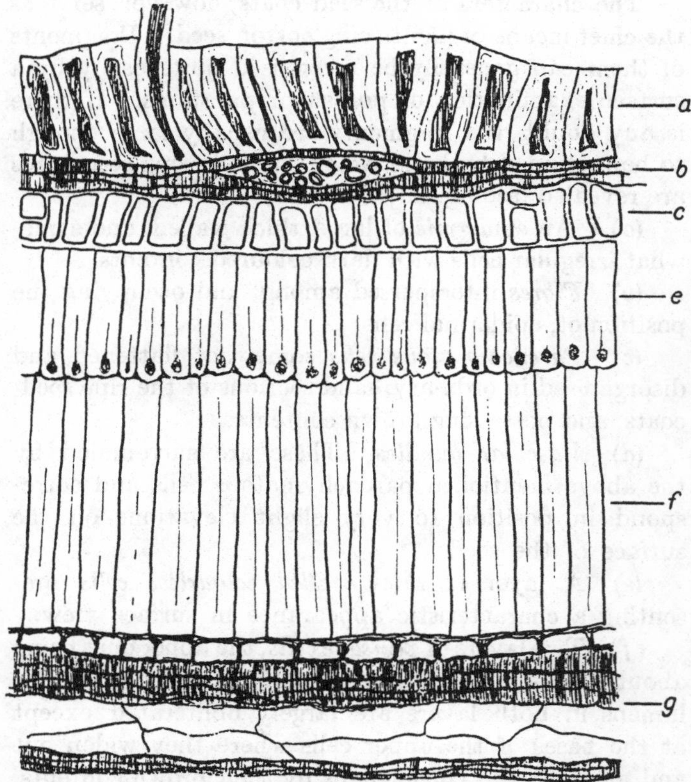

Fig. 38. *Cotton.* Transverse section of seed-coat. *a,* epidermal cells with base of one hair showing. *b,* brown parenchymatous cells enclosing a vascular bundle. *c,* thick walled colourless cells. *e* and *f,* upper and lower palisade layers. *g,* layers of parenchyma. Shaded parts indicate brown colouring matter. ×250.

the husk alone as they remain attached to the cotyledonary tissue.

Surface views of some of the layers are shown in fig. 39. The most important features in these are the long fibres, the irregular, thick walled, epidermal cells

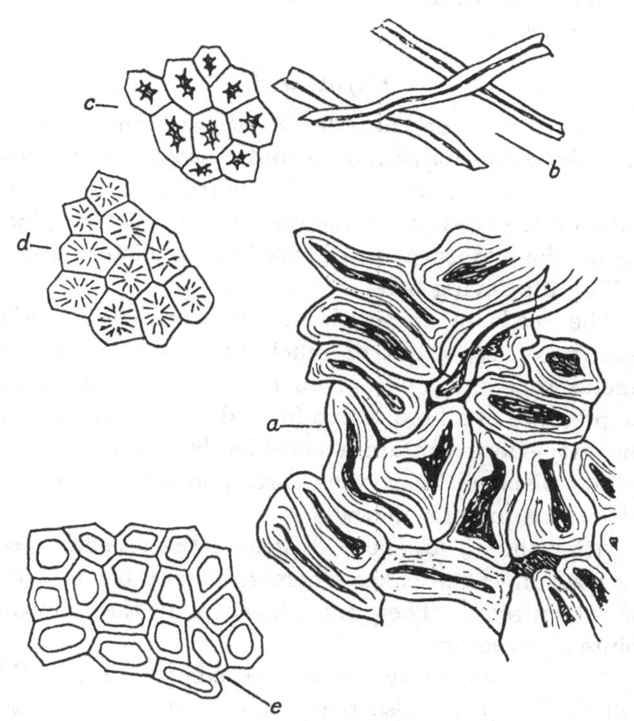

Fig. 39. *Cotton.* Surface preparation of seed-coat. *a*, epidermis with hair. *b*, short lengths of hairs. *c*, upper palisade cells—basal portion in focus. *d*, lower palisade cells. *e*, thick walled colourless cells. ×333.

arranged more or less in groups converging upon the point where a fibre grows out, and the thick-walled colourless cells in portions from which the epidermis is stripped off. The palisade cells are best seen at the edges of the preparation by squeezing the cover-glass. In surface view the palisade cells present different appearances according to the region in focus. [See fig. 39.]

Kapok seed.

Several trees of the order *Bombaceae*, among which are *Eriodendron anfructuosum* and *Bombas malabaricum*, furnish a fibre known as silky cotton. This is not borne on the seeds as in the closely allied cotton plant, but on the endocarp, and is used in upholstery and in making life-saving belts.

The seeds of these plants are somewhat smaller than cotton seed, but of much the same colour and general appearance except for the absence of hairs and the presence of a kind of cap formed from the funiculus. The cotyledons contain a considerable amount of oil, but no black resin cavities corresponding to those of cotton are present in them.

After the oil has been expressed the crushed seeds are used in cattle cakes where they may be mistaken for cotton seed. They differ however in the following points of structure:

1. The epidermal cells are comparatively thin walled. They are also polygonal in surface view and rectangular in section and do not bear hairs. [See figs. 40 and 41; cf. fig. 39.]

2. Gland-like depressions, one of which is shown in section in fig. 40 and in surface view in fig. 41, are distributed over the surface of the seed-coat.

The floor of a depression is made up of cells containing brown pigment, while the wall consists of a ring of colourless, thick walled, secretory cells supported by the cells of the outer brown coat.

3. No vascular bundles are present in the outer brown coat.

4. Three or four layers of colourless cells are

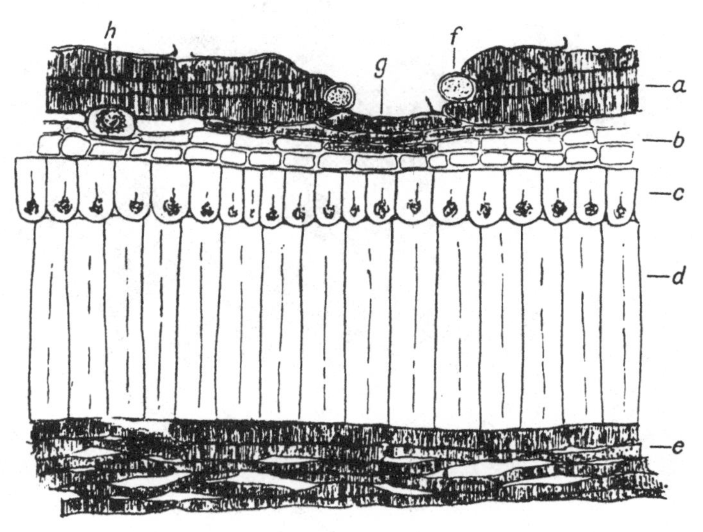

Fig. 40. *Kapok seed.* Transverse section of spermoderm. *a*, outer brown coat with cuticle covering epidermal cells. *b*, colourless cells. *c*, upper palisade cells. *d*, lower palisade cells. *e*, inner brown coat. *f*, colourless secretory cells cut across. *g*, brown cells forming floor of depression. *h*, crystal. The shaded portions represent cells containing brown pigment. × 200.

present instead of only one as in cotton. The outermost layer consists of thin walled cells with large intercellular spaces [*b*, fig. 41], while the innermost layer consists of fairly thick-walled cells compactly arranged [*c*, fig. 41]. The cell-walls of this layer are not however as thick as in the corresponding layer in cotton.

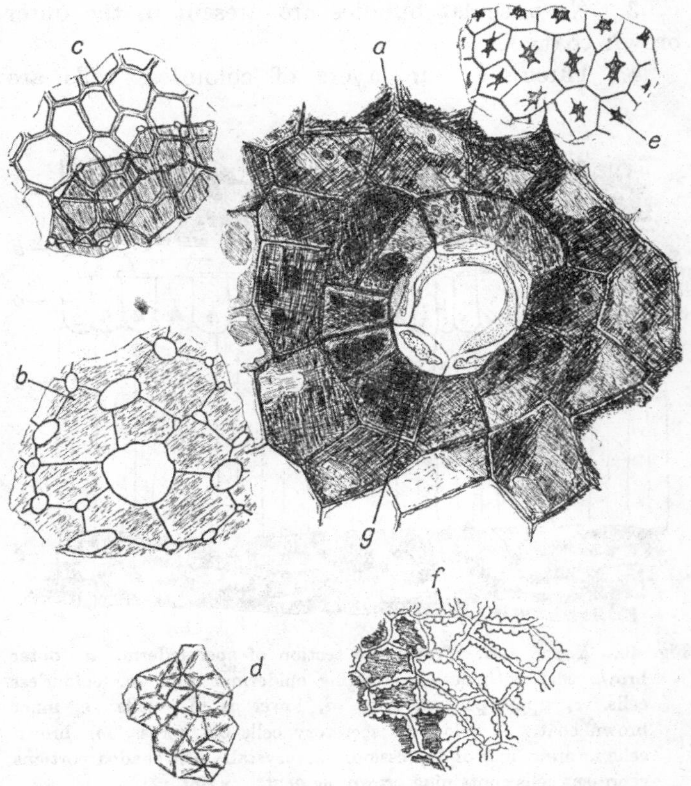

Fig. 41. *Kapok seed coat.* Surface preparation. *a,* epidermal cells. *b,* outer colourless cells. *c,* two innermost layers of colourless cells. *d,* upper surface of palisade cells. *e,* upper palisade cells with lumens in focus. *f,* perisperm. Figures 40 and 41 should be carefully compared with figs. 38 and 39 × 170.

5. The palisade cells, which cannot be distinguished from those of cotton in surface preparations, are seen to

be much shorter in section [compare figs. 38 and 40]. The upper ones are 30 μ as against 50 μ in cotton and the lower ones about 90 μ as against 110 μ.

6. The cells of the perisperm [*f*, fig. 41] in kapok are larger than in cotton, and have fewer projections on their walls.

8. No resin cavities are present in the cotyledons.

The two kinds of silky cotton seeds mentioned here resemble one another so closely that it is impossible to distinguish between them in ordinary surface preparations. Possibly they may differ in certain fine points, but these have not been worked out.

Linseed.

The seeds of the flax plant (*Linum usitatissimum* L.) like those of the cotton plant yield a valuable oil, and again the pressed material containing a certain amount of residual oil is sold in the form of cattle cakes. It also forms the basis for many calf meals and cream equivalents.

The seed itself is ovate and flattened, about 6 mm. in length, of a light brown colour, and possesses a highly polished surface.

In cross section the following layers are seen:

(i) *Epidermis* of fairly large cells containing mucilaginous material which at once exudes in the presence of water[1]. The outside of these cells is protected by a longitudinally striated cuticle.

(ii) Two layers of parenchymatous cells which

[1] The presence of this mucilaginous material makes it necessary that sections should be moistened with alcohol instead of water. As the seeds are small and hard it is difficult to get good hand sections.

from their appearance in surface view are known as *round cells*.

(iii) A layer of longitudinally elongated fibres with thick, pitted walls.

(iv) A layer of cross cells (transversely elongated).

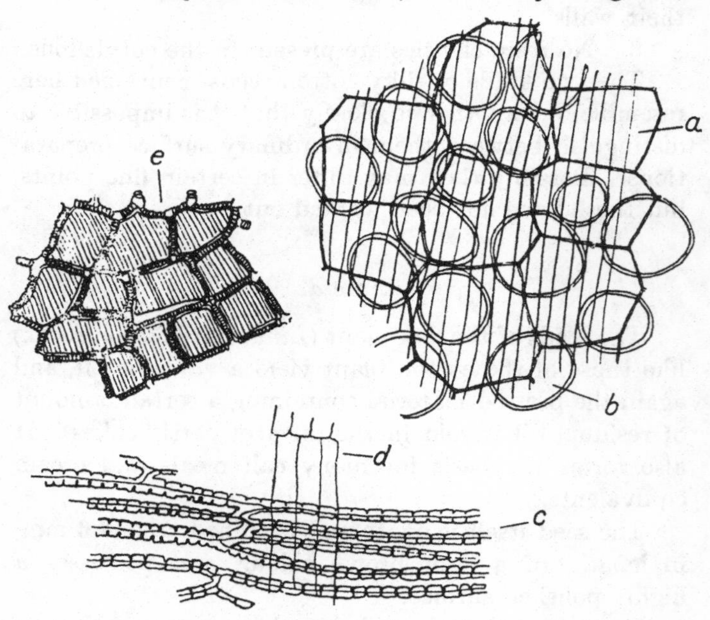

Fig. 42. *Linseed*. Surface preparation of seed-coat. *a*, epidermal cells cleared of mucilaginous contents. Striations on the cuticle indicated by the parallel lines. *b*, round cells. *c*, fibres. *d*, cross cells. *e*, pigment cells (brown). ×320.

(v) A layer of cells containing a deep-brown pigment—the *pigment layer*. The walls of these cells are distinctly pitted.

(vi) Endosperm, of several layers of parenchymatous cells containing much oil.

(vii) Cotyledons—also of oil-containing tissue.

Fig. 42 shows the above mentioned layers in surface view in a preparation which has been treated with water and cleared with potash. The epidermal cells are seen as thin walled cells devoid of mucilaginous contents. The striations of the cuticle are visible and one layer of round cells is shown somewhat more widely separated than usual. The fibres run in the same direction as the striations of the cuticle.

If much linseed is present in food material, it is advisable to get rid of the mucilage by boiling in dilute acid. The fine points of structure of the epidermis will be lost, but the other layers will stand out with sufficient clearness for identification purposes.

Niger seed.

The achenes of the Abyssinian plant *Guizotia oleifera*, one of the Compositae, are beginning to be used for feeding purposes after the bulk of their oil has been expressed.

These fruits are about 5 mm. in length and 1 mm. in breadth. They appear to be of a uniform greyish black colour, but closer examination with a lens shows that they are longitudinally striated, a very narrow light band alternating with a broader black band.

Fig. 38 represents a surface view of the fruit wall under low power. The alternating bands are clearly seen and a mottled appearance due to irregular distribution of pigment in the dark bands is also evident.

A cross section through the fruit wall is shown in fig. 44. The epidermis consists of longitudinally elongated, thin-walled cells. The hypodermal cells resemble the column cells of Leguminosae, and contain brown pigment. The pigment cells are dead black, and very striking in appearance. The fibres are

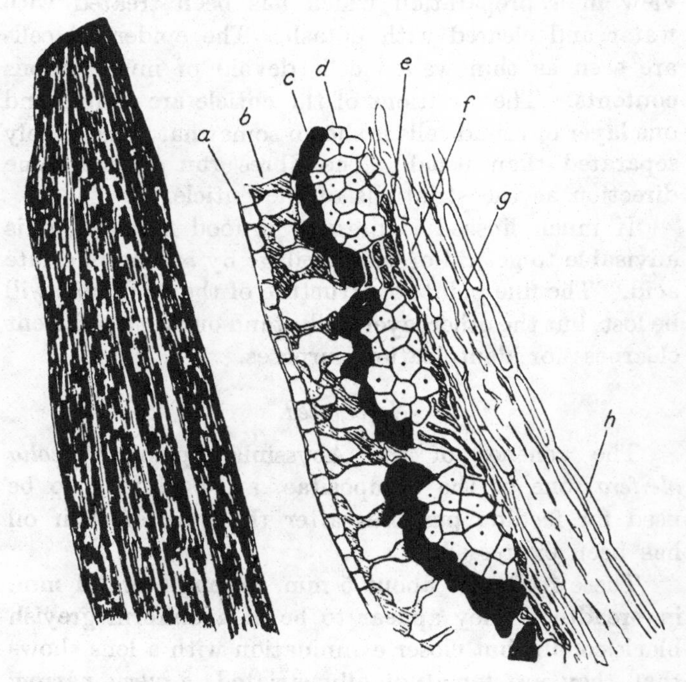

Fig. 43. *Niger*. Surface view of fruit wall. ×33.

Fig. 44. *Niger*. Transverse section of fruit wall. *a*, epidermis. *b*, hypodermal cells with brown contents. *c*, pigment cells. *d*, fibres. *e*, brown parenchyma. *f*, colourless parenchyma. *g*, spermoderm. *h*, aleurone layer. ×380.

characterised by the fact that their lumens are practically obliterated. The remaining layers are parenchyma with brown pigment, colourless parenchyma, spermoderm, and aleurone cells. Fig. 45 shows the somewhat characteristic spermoderm in surface view.

Note. The seed-coats of niger greatly resemble those of *Madia sativa*, another oil seed used in cakes, the difference being that in the latter the walls of the epidermis are beaded, the hypoderm cells are inconspicuous and not hour-glass or dumb-bell-shaped, and walls of the spermoderm straight and non-beaded.

Fig. 45. *Niger*. Spermoderm. Surface view. ×500.

Hemp.

The so-called hemp seed is really the fruit of the hemp plant (*Cannabis sativa* L.). It is an oval structure 4 mm. long by 3 mm. broad. The cotyledonary tissue which contains much oil is protected by a hard, grey pericarp inside of which is a delicate, dark green spermoderm, with a brown cap at one end. In some cases the dried calyx will be found covering the pericarp.

Histology.

(i) *Calyx.* Fragments of the calyx in surface view show an *outer epidermis* [*a*, fig. 48] with thick-walled

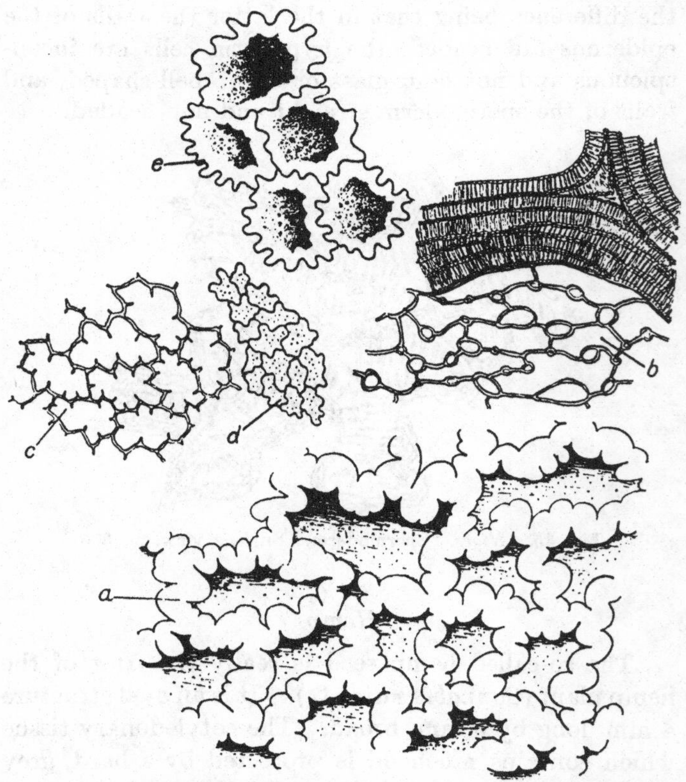

Fig. 46. *Hemp*. Surface preparation of fruit wall. *a*, epidermis. *b*, spongy tissue with vascular bundle. *c*, brown cells. *d*, dwarf cells. *e*, palisade cells. ×375.

unicellular hairs [each of which may contain one or more cystoliths in the enlarged basal part] and multicellular glandular hairs, a *mesophyll* of regular polygonal cells each containing a crystal mass of irregular shape [*d*, fig. 48], and an *inner epidermis* of thin-walled wavy cells not shown in the figure.

(ii) *Pericarp*. This consists of the following layers:

(*a*) An *epidermis* of cells of which the radial walls are thick and wavy, and the tangential ones pitted [*a*, fig. 46].

(*b*) *Spongy tissue* [*b*, fig. 46]. This is not always easy to see in surface preparations as it is apt to be masked by the more conspicuous layers which underly it. It contains numerous branching vascular strands.

(*c*) *Brown parenchymatous cells* [*c*, fig. 46]. These have projections on their walls which may almost divide the cells up into compartments.

(*d*) *Dwarf-cells* with pitted tangential walls [*d*, fig. 46].

(*e*) *Palisade cells* [*e*, fig. 46]. These form the most conspicuous layer of the fruit wall. They may be 100 μ in height by 30 μ or 40 μ wide. Their radial walls are seen to be wavy both in surface view and in cross section, and are so thick that the lumens of the cells are practically obliterated except near the base.

(iii) *Spermoderm*. Two distinct layers can be made out in the spermoderm, viz.:

(1) A layer of elongated cells between which are rows of intercellular spaces [*a*, fig. 47].

(2) A spongy parenchyma underlying the above, in which it is difficult to make out the outlines of the cells. It consists of several layers of flattened cells, some of which appear to be star-shaped, while in others the intercellular spaces are oval or circular.

Both layers of the spermoderm contain green colouring matter which is insoluble in alcohol, ether, or alkali.

The *endosperm* consists of a layer of aleurone cells similar to those found in cereals. It is one cell thick

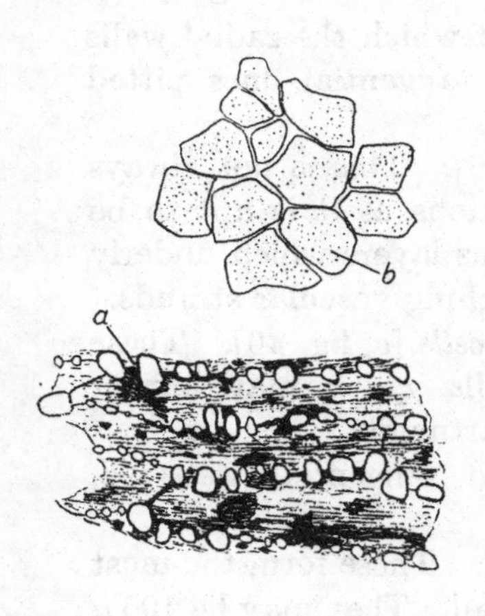

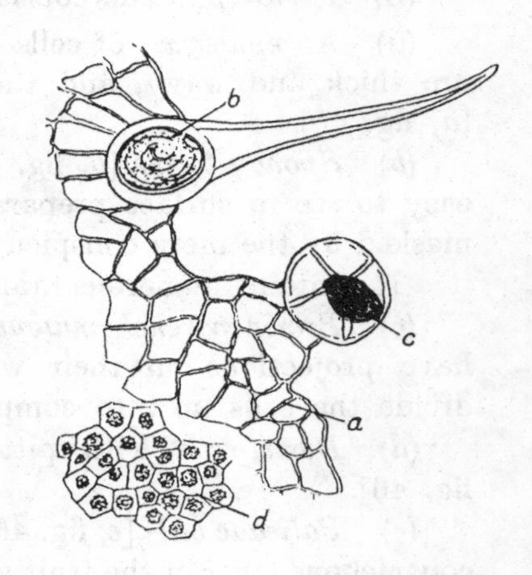

Fig. 47. *Hemp.* Spermoderm and endosperm. *a*, elongated cells of spermoderm. *b*, aleurone cells

Fig. 48. *Hemp.* Calyx. *a*, epidermis (outer). *b*, cystolith. *c*, glandular hair. *d*, crystal cells.

except in the region beneath the radicle where there are several layers of cells. [See fig. 47 *b*.]

A *perisperm* consisting of delicate elongated cells is present outside the endosperm.

CRUCIFEROUS SEEDS.

The cruciferous seed residues found in mixed cattle cakes are derived almost entirely from plants of the genus Brassica.

These seeds are mostly small and spherical, of a yellow, yellow brown, dark brown or black colour, and possess little or no polish. They can usually be picked out with fair certainty by the naked eye from seeds belonging to other families in material which has been boiled in potash, but it is not an easy matter to identify the different species of this group.

Typically, a seed-coat of one of the Brassicas consists of the following layers from without inwards:

(*a*) *An epidermis* of rather large, mucilaginous cells.

(*b*) *A single or double sub-epidermal layer* which varies in its nature in different seeds or may even be lost in the ripe seed.

(*c*) *A layer of palisade cells.* These may be all the same length or they may vary in length giving rise to a meshwork appearance in surface view, the dark coloured portions being in the region of the long cells and the lighter portions in the region of the short cells. Each space between a group of short palisade cells and the epidermal cells is occupied by single sub-epidermal cells known as giant cells, while the surrounding long palisade cells reach to the bases of the epidermal cells.

(*d*) *Parenchyma*—two or three layers of thin walled cells.

(*e*) *Aleurone* cells.

The principal seeds to consider are white mustard, black mustard, charlock, and rape.

White mustard. (Brassica alba or Sinapis alba.)

This plant has seeds 2 or 3 mm. in diameter, spherical and buff-coloured. The histological features of the seed-coats are as follows:

(*a*) *Epidermis* of polygonal mucilage cells about 100 μ in diameter. In water the mucilage exudes and the cells tend to disorganise, so that preparations should be made by mounting in alcohol instead of water. The lumens of the cells are seen to be very narrow, and the mucilaginous substance is arranged in concentric layers [*a*, fig. 41].

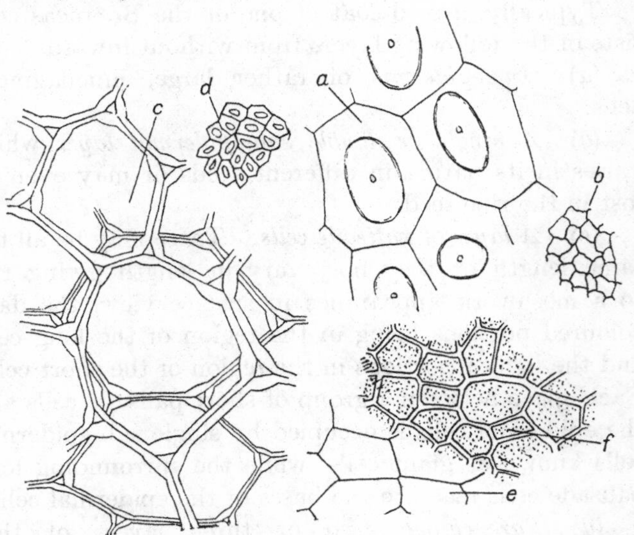

Fig. 49. *White mustard.* Surface preparation. *a*, epidermal cells. *b*, outer part of palisade cells in focus (so-called reticulum). *c*, palisade cells—basal part in focus. *d*, hypodermal cells—two layers. *e*, parenchyma. *f*, aleurone cells. ×250.

(*b*) *The sub-epidermal layer* which in white mustard is very characteristic. It consists of two layers of thick walled, hexagonal, collenchymatous cells, 100 μ or more in diameter, and superimposed one upon the other [*c*, fig. 49].

(*c*) *Palisade cells.* These are radially elongated, and of nearly uniform height, hence the absence of reticulations in this seed. The radial walls are thick at the bases of cells, but thin out towards the outer surface. Hence in surface preparations the outer part of the palisade layer when in focus appears as a fine reticulum [*b*, fig. 49] and the basal or inner part appears to consist of thick-walled cells with small lumens [*d*, fig. 49]. The walls of these cells differ from those of most of the cruciferous seeds in being colourless.

The *parenchymatous* and *aleurone layers* are figured at *e* and *f*, fig. 49, respectively.

Black mustard. (Brassica nigra L.)

The reticulated appearance of these seeds due to the cause mentioned on p. 63 is represented in fig. 50.

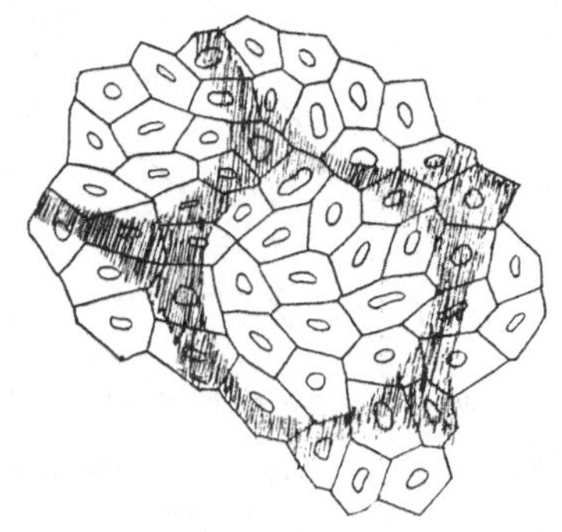

Fig. 50. *Black mustard.* Surface preparation of palisade layer showing reticulated appearance due to unequal height of cells. ×250.

The pigmented palisade cells only are represented in the figure (they form by far the most conspicuous layer in these seeds). The epidermal cells however resemble those of white mustard; the sub-epidermal cells are of the giant cell type described above (one giant cell overlies each of the light portions in the figure), and the parenchyma and aleurone layers are similar to those of white mustard except that the parenchyma is pigmented.

It is very difficult to distinguish between the seed

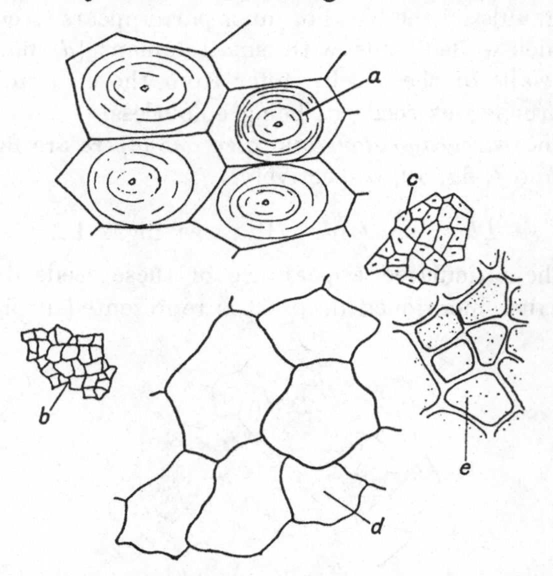

Fig. 51. *Charlock.* Surface preparation of seed-coat. *a*, epidermal cells with mucilaginous contents. *b*, reticulum (the thin walled outer portion of the palisade cells). *c*, basal part of palisade cells in focus. *d*, parenchyma. *e*, aleurone cells. ×250.

coats of black mustard (*Brassica nigra*) and those of brown mustard (*Brassica Besseriana*), sometimes called Sarepta mustard. According to Winton the palisade cells of the latter seed are 'somewhat wider (often triangular) and the sub-epidermal layer is much less distinct, often being entirely obliterated.'

Charlock. (Sinapis arvensis L. *or* Brassica arvensis L.)

This differs from black mustard in having no reticulations, and no apparent sub-epidermal layer in the ripe seed; the outer portion of the palisade layer is thin walled as in white mustard.

For the seed-coat of charlock see fig. 51. Charlock is also distinguished by the fact that its seed-coats turn blood red in colour when they are heated with acidified chloral hydrate [Winton].

Rape.

Here the epidermis and the sub-epidermal layers are not apparent in the ripe seed.

The palisade and other layers are similar to those of charlock; they do not, however, give the blood red reaction with acid chloral hydrate.

Rape is the commonest cruciferous seed used in cattle-foods, but the reticulated seed-coats of black mustard are also quite common.

Table of Cruciferous Seeds.

Husks yellow no reticulations, epidermal and sub-epidermal layers structureless	Indian colza
Husks yellow, no reticulations, hexagonal collenchymatous sub-epidermal cells	White mustard
[Husks yellow, no collenchyma, reticulations present	? Unripe seed of black mustard]
Husks dark coloured, no reticulations	Rape or charlock
Husks dark coloured, reticulated	Black mustard

NUTS.

1. *Palm nuts.*

The so-called palm nuts are the seeds of the Oil Palm (*Elaeis Guineensis* L.) and are enclosed in thick, hard endocarps. Palm nut cake or palm nut meal consists of these seeds in a finely ground condition with the oil expressed and most of the endocarp removed.

Fragments of palm nut cake or meal on being cleared with potash and examined microscopically are seen to consist of:

(*a*) Masses of endosperm cells with curiously thickened walls [see fig. 52]. These cells form the main bulk of the meal, and have largely lost their contents which originally consisted of fat and aleurone grains.

(*b*) Brown masses of seed-coat which appear structureless owing to their thickness and opacity. They really consist of layers of regular flattened cells with brown contents.

(*c*) Masses of rounded or irregularly shaped stone cells from the endocarp. [Fig. 52.]

2. *Coco Nuts.* (Cocos nucifera.)

These are similar morphologically to palm nuts but are much larger.

Examination of the meal reveals brown masses of seed-coat not distinguishable from those in the palm nut meal.

The endosperm cells however are much longer and narrower than those of palm nut and have thin non-beaded walls. As in the former case the contents of the cells will be found largely to have disappeared. For palm nut and coco nut see figs. 52 and 53.

Palm nuts and coco nuts are both made into feeding cakes of which they are the sole ingredients. They also find their way into mixed cattle cakes, pig meals and calf meals.

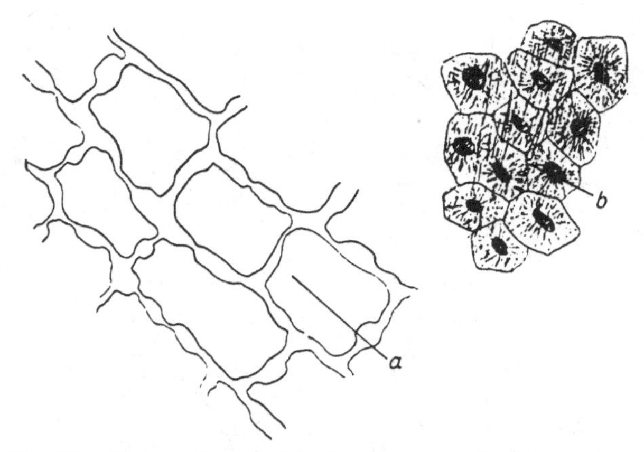

Fig. 52. *Palm nut.* *a*, cells of endosperm. *b*, stone cells of endocarp (brown). ×250.

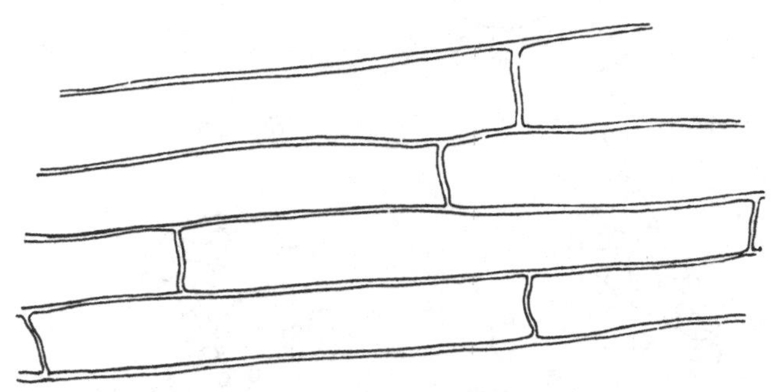

Fig. 53. *Coco nut.* Cells of endosperm. ×250.

As far as feeding value goes the nut cakes are considered to be the poorest of the feeding cakes.

APPENDIX

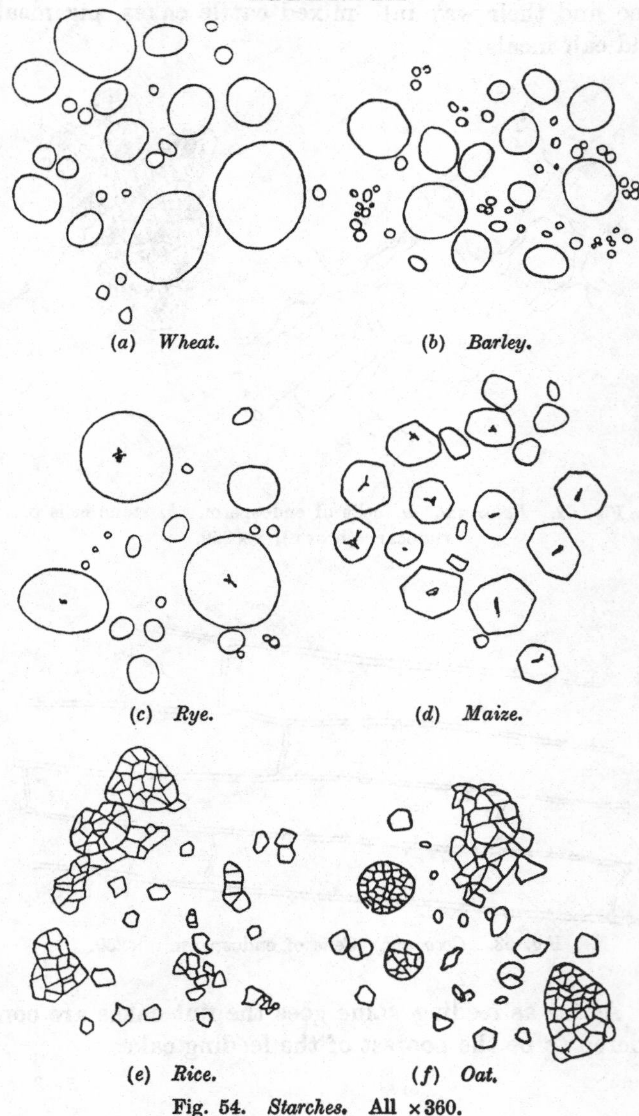

Fig. 54. *Starches.* All ×360.

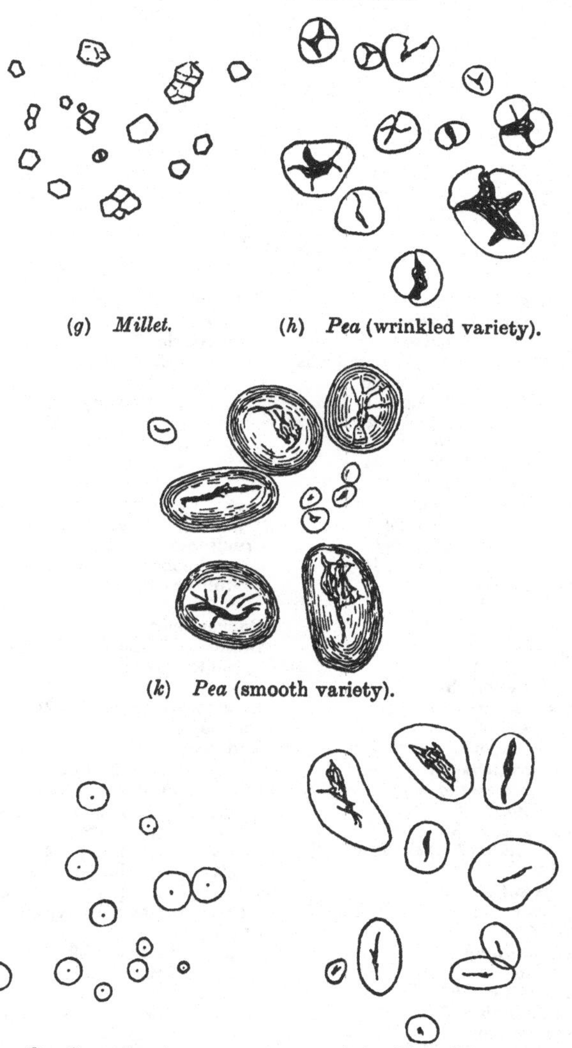

(*g*) *Millet.* (*h*) *Pea* (wrinkled variety).

(*k*) *Pea* (smooth variety).

(*l*) *Ground nut.* (*m*) *Horse bean.*

Fig. 54. *Starches.* All ×360.

These figures are referred to in the text, and are grouped together here for the sake of comparison.

INDEX

Printed in the United States
By Bookmasters